Lecture Notes in Mathematics   2136

More information about this series at http://www.springer.com/series/304

Valerio Capraro • Martino Lupini

Introduction to Sofic and Hyperlinear Groups and Connes' Embedding Conjecture

With an Appendix by Vladimir Pestov

 Springer

Valerio Capraro
Center for Mathematics and Computer
    Science (CWI)
Amsterdam
The Netherlands

Martino Lupini
Department of Mathematics
California Institute of Technology
Pasadena
CA, USA

ISSN 0075-8434   ISSN 1617-9692   (electronic)
Lecture Notes in Mathematics
ISBN 978-3-319-19332-8   ISBN 978-3-319-19333-5   (eBook)
DOI 10.1007/978-3-319-19333-5

Library of Congress Control Number: 2015945815

Mathematics Subject Classification (2010): 20F65, 20F69, 03C20, 03C98, 46L10, 46M07

Springer International Publishing AG Switzerland is part of Springer Science+Business Media
(www.springer.com)

Preface

Analogy is one of the most effective techniques of human reasoning: When we face new problems, we compare them with simpler and already known ones, in the attempt to use what we know about the latter ones to solve the former ones. This strategy is particularly common in Mathematics, which offers several examples of abstract and seemingly intractable objects: Subsets of the plane can be enormously complicated but, as soon as they can be approximated by rectangles, then they can be measured; Uniformly finite metric spaces can be difficult to describe and understand but, as soon as they can be approximated by Hilbert spaces, then they can be proved to satisfy the coarse Novikov's and Baum-Connes's conjectures.

These notes deal with two particular instances of such a strategy: Sofic and hyperlinear groups are in fact the countable discrete groups that can be approximated in a suitable sense by finite symmetric groups and groups of unitary matrices. These notions, introduced by Gromov and Rădulescu, respectively, at the end of the 1990s, turned out to be very deep and fruitful, and stimulated in the last 15 years an impressive amount of research touching several seemingly distant areas of mathematics including geometric group theory, operator algebras, dynamical systems, graph theory, and more recently even quantum information theory. Several long-standing conjectures that are still open for arbitrary groups were settled in the case of sofic or hyperlinear groups. These achievements aroused the interest of an increasing number of researchers into some fundamental questions about the nature of these approximation properties. Many of such problems are to this day still open such as, outstandingly: Is there any countable discrete group that is not sofic or hyperlinear? A similar pattern can be found in the study of II_1 factors. In this case, the famous conjecture due to Connes (commonly known as Connes' embedding conjecture) that any II_1 factor can be approximated in a suitable sense by matrix algebras inspired several breakthroughs in the understanding of II_1 factors, and stands out today as one of the major open problems in the field.

The aim of this monograph is to present in a uniform and accessible way some cornerstone results in the study of sofic and hyperlinear groups and Connes' embedding conjecture. These notions, as well as the proofs of many results, are here presented in the framework of model theory for metric structures. We believe that this point of view, even though rarely explicitly adopted in the literature, can contribute to a better understanding of the ideas therein, as well as provide additional tools to attack many remaining open problems. The presentation is nonetheless self-contained and accessible to any student or researcher with a graduate-level mathematical background. In particular, no specific knowledge of logic or model theory is required.

Chapter 1 presents the conjectures and open problems that will serve as common thread and motivation for the rest of the survey: Connes' embedding conjecture, Gottschalk's conjecture, and Kaplansky's conjecture. Chapter 2 introduces sofic and hyperlinear groups, as well as the general notion of metric approximation property; outlines the proofs of Kaplansky's direct finiteness conjecture and the algebraic eigenvalues conjecture for sofic groups; and develops the theory of entropy for sofic group actions, yielding a proof of Gottschalk's surjunctivity conjecture in the sofic case. Chapter 3 discusses the relationship between hyperlinear groups and the Connes' embedding conjecture; establishes several equivalent reformulations of the Connes' embedding conjecture due to Haagerup-Winsløw and Kirchberg; describes the purely algebraic approach initiated by Rădulescu and carried over by Klep-Schweighofer and Juschenko-Popovich; and finally outlines the theory of Brown's invariants for II$_1$ factors satisfying the Connes' embedding conjecture. An appendix by V. Pestov provides a pedagogically new introduction to the concepts of ultrafilters, ultralimits, and ultraproducts for those mathematicians who are not familiar with them, and aiming to make these concepts appear very natural.

The choice of topics is unavoidably not exhaustive. A more detailed introduction to the basic results about sofic and hyperlinear groups can be found in [125, 126]. The surveys [118, 120, 121] contain several other equivalent reformulations of the Connes' embedding conjecture in purely algebraic or C*-algebraic terms.

This survey originated from a short intensive course that the authors gave at the Universidade Federal de Santa Catarina in 2013 in occasion of the "Workshop on sofic and hyperlinear groups and the Connes' embedding conjecture" supported by CAPES (Brazil) through the program "Science without borders", PVE project 085/2012. We would like to gratefully thank CAPES for its support, as well as the organizers of the workshop Daniel Gonçalves and Vladimir Pestov for their kind hospitality, and for their constant and passionate encouragement.

Moreover, we are grateful to Hiroshi Ando, Goulnara Arzhantseva, Samuel Coskey, Ilijas Farah, Tobias Fritz, Benjamin Hayes, Liviu Păunescu, Vladimir Pestov, David Sherman, Alain Valette, and five anonymous referees for several useful comments and suggestions.

Amsterdam, The Netherlands   Valerio Capraro
CA, USA   Martino Lupini

Contents

Chapter 1
Introduction

Valerio Capraro and Martino Lupini

1.1 von Neumann Algebras and II$_1$ Factors

Denote by $B(H)$ the algebra of bounded linear operators on the Hilbert space H. Recall that $B(H)$ is naturally endowed with an involution $x \mapsto x^*$ associating with an operator x its *adjoint* x^*. The *operator norm* $\|x\|$ of an element of $B(H)$ is defined by

$$\|x\| = \sup \{\|x\xi\| : \xi \in H, \ \|\xi\| \le 1\}.$$

Endowed with this norm, $B(H)$ is a Banach algebra with involution satisfying the identity

$$\|x^*x\| = \|x\|^2 \qquad \text{(C*-identity)}$$

i.e. a *C*-algebra*.

The *weak operator topology* on $B(H)$ is the weakest topology making the map

$$x \mapsto \langle x\xi, \eta \rangle$$

V. Capraro (✉)
Center for Mathematics and Computer Science (CWI), Amsterdam, The Netherlands
e-mail: caprarovalerio@gmail.com

M. Lupini
Department of Mathematics, California Institute of Technology, 1200 East California Boulevard, Pasadena, CA 91125, USA
e-mail: mlupini@mathstat.yorku.ca

© Springer International Publishing Switzerland 2015
V. Capraro, M. Lupini, *Introduction to Sofic and Hyperlinear Groups and Connes' Embedding Conjecture*, Lecture Notes in Mathematics 2136, DOI 10.1007/978-3-319-19333-5_1

continuous for every $\xi, \eta \in H$, where $\langle \cdot, \cdot \rangle$ denotes the scalar product of H. The *strong operator topology* on $B(H)$ is instead the weakest topology making the maps

$$x \mapsto \|x\xi\|$$

continuous for every $\xi \in H$. As the names suggest the strong operator topology is stronger than the weak operator topology. It is a consequence of the Hahn-Banach theorem that, conversely, a convex subset of $B(H)$ closed in the strong operator topology is also closed in the weak operator topology (see Theorem 5.1.2 of [91]).

A (concrete) *von Neumann algebra* is a unital *-subalgebra (i.e. closed with respect to taking adjoints) of $B(H)$ that is closed in the weak (or, equivalently, strong) operator topology. It is easy to see that if X is a subset of $B(H)$, then the intersection of all von Neumann algebras $M \subset B(H)$ containing X is again a von Neumann algebra, called the von Neumann algebra generated by X. Theorem 1.1.1 is a cornerstone result of von Neumann, known as von Neumann double commutant theorem, asserting that the von Neumann algebra generated by a subset X of $B(H)$ can be characterized in a purely algebraic way. The *commutant X'* of a subset X of $B(H)$ is the set of $y \in B(H)$ commuting with every element of X. The *double commutant X''* of X is just the commutant of X'.

Theorem 1.1.1 *The von Neumann algebra generated by a subset X of $B(H)$ containing the unit and closed with respect to taking adjoints coincides with the double commutant X'' of X.*

A *faithful normal trace* on a von Neumann algebra M is a linear functional τ on M such that:

- $\tau(x^*x) \geq 0$ for every $x \in M$ (τ is *positive*);
- $\tau(x^*x) = 0$ implies $x = 0$ (τ is *faithful*);
- $\tau(xy) = \tau(yx)$ for every $x, y \in M$ (τ is *tracial*);
- $\tau(1) = 1$ (τ is *unital*)
- τ is continuous on the unit ball of M with respect to the weak operator topology (τ is *normal*).

A (finite) von Neumann algebra endowed with a distinguished trace will be called a *tracial von Neumann algebra*. A tracial von Neumann algebra is always *finite* as in [92, Definition 6.3.1]. Conversely any finite von Neumann algebra faithfully represented on a separable Hilbert space has a faithful normal trace by Kadison and Ringrose [92, Theorem 8.2.8].

The *center* $Z(M)$ of a von Neumann algebra $M \subset B(H)$ is the subalgebra of M consisting of the operators in M commuting with any other element of M. A von Neumann algebra M is called a *factor* if its center is as small as possible, i.e. it contains only the scalar multiples of the identity. A finite factor has a unique faithful normal trace (see [92, Theorem 8.2.8]). Moreover any faithful unital tracial positive linear functional on a finite factor is automatically normal, and hence coincides with its unique faithful normal trace.

Example 1.1.2 If H_n is a Hilbert space of finite dimension n, then $B(H_n)$ is a finite factor denoted by $M_n(\mathbb{C})$ isomorphic to the algebra of $n \times n$ matrices with complex coefficients. The unique trace on $M_n(\mathbb{C})$ is the usual normalized trace of matrices.

It is a consequence of the type classification of finite factors (see [92, Sect. 6.5]) that the ones described in Example 1.1.2 are the unique examples of finite factors that are finite dimensional as vector spaces.

Definition 1.1.3 A II_1 *factor* is an infinite-dimensional finite factor.

Theorem 1.1.4 is a cornerstone result of Murray and von Neumann (see [115, Theorem XIII]), offering a characterization of II_1 factors within the class of finite factors.

Theorem 1.1.4 *If M is a II_1 factor, then M contains, for every natural number n, a unital copy of $M_n(\mathbb{C})$, i.e. there is a trace preserving *-homomorphism from $M_n(\mathbb{C})$ to M.*

It follows from weak continuity of the trace and the type classification of finite factors that if M is a finite factor, then the following statements are equivalent:

1. M is a II_1 factor;
2. the trace τ of M attains on projections all the real values between 0 and 1.

The unique trace τ on a finite factor M allows to define the Hilbert-Schmidt norm $\|\cdot\|_2$ on M, by $\|x\|_2 = \tau(x^*x)^{\frac{1}{2}}$. The Hilbert-Schmidt norm is continuous with respect to the operator norm $\|\cdot\|$ inherited from $B(H)$. A finite factor is called *separable* if it is separable with respect to the topology induced by the Hilbert-Schmidt norm.

Let us now describe one of the most important constructions of II_1 factors. If Γ is a countable discrete group then the *complex group algebra* $\mathbb{C}\Gamma$ is the complex algebra of formal finite linear combinations

$$\lambda_1 \gamma_1 + \cdots + \lambda_k \gamma_k$$

of elements of γ with coefficients from $\mathbb{C}$. Any element of $\mathbb{C}\Gamma$ can be written as

$$\sum_\gamma a_\gamma \gamma,$$

where $(a)_{\gamma \in \Gamma}$ is a family of complex numbers all but finitely many of which are zero. Sum and multiplication of elements of $\mathbb{C}\Gamma$ are defined by

$$\left(\sum_\gamma a_\gamma \gamma \right) + \left(\sum_\gamma b_\gamma \gamma \right) = \sum_\gamma \left(a_\gamma + b_\gamma \right) \gamma$$

and

$$\left(\sum_\gamma a_\gamma \gamma\right)\left(\sum_\gamma b_\gamma \gamma\right) = \sum_\gamma \left(\sum_{\rho\rho'=\gamma} a_\rho b_{\rho'}\right)\gamma.$$

Consider now the Hilbert space $\ell^2(\Gamma)$ of square-summable complex-valued functions on Γ. Each $\gamma \in \Gamma$ defines a unitary operator λ_γ on $\ell^2(\Gamma)$ by:

$$\lambda_\gamma(f)(x) = f(\gamma^{-1}x).$$

The function $\gamma \to \lambda_\gamma$ extends by linearity to an embedding of $\mathbb{C}\Gamma$ into the algebra $B\left(\ell^2(\Gamma)\right)$ of bounded linear operators on $\ell^2(\Gamma)$.

Definition 1.1.5 The weak closure of $\mathbb{C}\Gamma$ (identified with a subalgebra of $B\left(\ell^2(\Gamma)\right)$) is a von Neumann algebra denoted by $L\Gamma$ and called the *group von Neumann algebra* of Γ.

The group von Neumann algebra $L\Gamma$ is canonically endowed with the trace τ obtained by extending by continuity the condition:

$$\tau\left(\sum_\gamma a_\gamma \gamma\right) = a_{1_\Gamma},$$

for every element $\sum_\gamma a_\gamma \gamma$ of $\mathbb{C}\Gamma$.

Exercise 1.1.6 Assume that Γ is an ICC group, i.e. every nontrivial conjugacy class of Γ is infinite. Show that $L\Gamma$ is a II_1 factor.

Exercise 1.1.7 Show that the free group $\mathbb{F}_n$ on $n \geq 2$ generators and the group S_∞^{fin} of finitely supported permutations of a countable set are ICC.

It is a milestone result of Murray and von Neumann from [115] (see also Theorem 6.7.8 of [92]) that the group factors associated with the free group $\mathbb{F}_2$ and, respectively, the group S_∞^{fin} are nonisomorphic. It is currently a major open problem in the theory of II_1 factors to determine whether free groups over different number of generators have isomorphic associated factors.

Connes' embedding conjecture asserts that any separable II_1 factor can be approximated by finite dimensional factors, i.e. matrix algebras. More precisely:

Conjecture 1.1.8 (Connes [43]) If M is a separable II_1 factor with trace τ_M then for every $\varepsilon > 0$ and every finite subset F of M there is a function Φ from M to a matrix algebra $M_n(\mathbb{C})$ that on F approximately preserves the operations and the trace. This means that $\Phi(1) = 1$ and for every $x, y \in F$:

- $\|\Phi(x+y) - (\Phi(x) + \Phi(y))\|_2 < \varepsilon$;
- $\|\Phi(xy) - \Phi(x)\Phi(y)\|_2 < \varepsilon$;
- $\left|\tau_M(x) - \tau_{M_n(\mathbb{C})}(\Phi(x))\right| < \varepsilon$.

1.2   Voiculescu's Free Entropy

Connes' embedding conjecture is related to many other problems in pure and applied mathematics and can be stated in several different equivalent ways. The simplest and most practical way is probably through Voiculescu's free entropy.

Consider a random variable which outcomes the set $\{1, \ldots, n\}$ with probabilities $p_1, \ldots, p_n$. Observe that its Shannon's entropy, $-\sum p_i \log(p_i)$, can also be constructed through the following procedure:

1. Call *microstate* any function f from $\{1, 2, \ldots, N\}$ to $\{1, 2, \ldots, n\}$. A microstate f ε-*approximates* the discrete distribution $p_1, \ldots p_n$ if for every $i \in \{1, 2, \ldots, n\}$

$$\left| \frac{|f^{-1}(i)|}{N} - p_i \right| < \varepsilon.$$

   Denote the number of such microstates by $\Gamma(p_1, \ldots, p_n, N, \varepsilon)$.
2. Take the limit of

$$N^{-1} \log |\Gamma(p_1, \ldots, p_n, N, \varepsilon)|,$$

   as $N \to \infty$
3. Finally take the limit as ε goes to zero.

One can show that the result is just the opposite of the Shannon entropy, that is, $\sum_{i \in n} p_i \log p_i$. Over the early 1990s, Voiculescu, in part motivated by the isomorphism problem of whether the group von Neumann algebras associated to different free groups are isomorphic or not, realized that this construction could be imitated in the noncommutative world of II_1 factors.

1. Microstates are self-adjoint matrices, instead of functions between finite sets. Formally, let $\varepsilon, R > 0$, $m, k \in \mathbb{N}$, and $X_1, \ldots, X_n$ be free random variables[1] on a II_1-factor M. We denote $\Gamma_R(X_1, \ldots, X_n; m, k, \varepsilon)$ the set of $(A_1, \ldots, A_n) \in (M_k(\mathbb{C})_{sa})^n$ such that $\|A_m\| \leq R$ and

$$|tr(A_{i_1} \cdots A_{i_m}) - \tau(X_{i_1} \cdots X_{i_m})| < \varepsilon$$

   for every $1 \leq m \leq p$ and $(i_1, \ldots, i_m) \in \{1, 2, \ldots, n\}^m$.
2. The discrete measure is replaced by the Lebesgue measure λ on $(M_k(\mathbb{C})_{sa})^n$.
3. The limits are replaced by suitable limsups, sups, and infs.

---

[1]Freeness is the analogue of independence in the noncommutative framework of II_1-factors. For a formal definition we refer the reader to [147].

Specifically, set

$$\chi_R(X_1, \ldots, X_n; m, k, \varepsilon) := \log \lambda(\Gamma_R(X_1, \ldots X_n; m, k, \varepsilon)),$$

$$\chi_R(X_1, \ldots X_n; m, \varepsilon) := \limsup_{k \to \infty} (k^{-2} \chi_R(X_1, \ldots X_n; m, k, \varepsilon) + 2^{-1} n \log(k)),$$

$$\chi_R(X_1, \ldots X_n) := \inf\{\chi_R(X_1, \ldots X_n; m, \varepsilon) : m \in \mathbb{N}, \varepsilon > 0\},$$

Finally, define **entropy** of the variables $X_1, \ldots, X_n$ the quantity

$$\chi(X_1, \ldots X_n) := \sup\{\chi_R(X_1, \ldots X_n) : R > 0\}.$$

The factor k^{-2} instead of k^{-1} comes from the normalization. The addend $2^{-1} n \log(k)$ is necessary, since otherwise $\chi_R(X_1, \ldots X_n; m, \varepsilon)$ would always be equal to $-\infty$.

It is not clear why this construction should give entropy different from $-\infty$. Indeed Voiculescu himself proved that it is $-\infty$ when $X_1, \ldots, X_n$ are linearly dependent (see [148, Proposition 3.6]). A necessary condition to have $\chi(X_1, \ldots, X_n) > -\infty$ is that $\Gamma_R(X_1, \ldots, X_n, m, k, \varepsilon)$ is not empty for some k, i.e. the finite subset $X = \{X_1, \ldots, X_n\}$ of M_{sa} has microstates. This requirement turns to be equivalent to the fact that M satisfies Connes' embedding conjecture.

Theorem *Let M be a II_1 factor. The following conditions are equivalent*

1. *Every finite subsets $X \subseteq M_{sa}$ has microstates.*
2. *M verifies Connes' embedding conjecture.*

1.3 History of Connes' Embedding Conjecture

Connes' embedding conjecture had its origin in Connes' sentence: "We now construct an approximate imbedding of N in R. Apparently such an imbedding ought to exist for all II_1 factors because it does for the regular representation of free groups. However the construction below relies on condition 6" (see [43, page 105]). This seemingly innocent observation received in the first 15 years after having been formulated relatively little attention. The situation changed drastically in the early 1990s, when two fundamental papers appeared on the scene: the aforementioned [148] where the Connes embedding conjecture is used to define free entropy, and [101] where Kirchberg obtained several unexpected reformulations of the Connes embedding conjecture very far from its original statement, such as the fact that the maximal and minimal tensor products of the group C*-algebra of the free group on infinitely many generators coincide:

$$C^*(F) \otimes_{\min} C^*(F) = C^*(F) \otimes_{\max} C^*(F). \tag{1.1}$$

What is most striking in this equivalence is that Connes' embedding conjecture concerns the class of all separable II_1 factors, while (1.1) is a statement about a single C*-algebra. Even more surprising is the fact that the equivalence of these statements can be proved a topological way as shown in [83].

Following the aforementioned papers by Voiculescu and Kirchberg, a series of papers from different authors proving various equivalent reformulations of Connes' embedding conjecture appeared, such as [16, 41, 64, 84, 118] contributing to arouse the interest around this conjecture. In particular Florin Rădulescu showed in [130] that Connes' embedding conjecture is equivalent to some noncommutative analogue of Hilbert's 17th problem. This in turn inspired work of Klep and Schweighofer, who proved in [103] a purely algebraic reformulation of Connes' embedding conjecture (see also the more recent work [89]). In [25] it is observed that Connes' embedding conjecture could theoretically be checked by an algorithm if a certain problem of embedding Hilbert spaces with some additional structure into the Hilbert space $L^2(M, \tau)$ associated to a II_1 factor has a positive solution. Another computability-theoretical reformulation of the Connes' embedding conjecture has more recently been proved in [73]. Other very recent discoveries include the fact that Connes' embedding conjecture is related to Tsirelson's problem, a major open problem in Quantum Information Theory (see [66, 88, 121]) and that it is connected to the "minimal" and "commuting" tensor products [97] of some group operator systems [35, 62, 63].

1.4 Hyperlinear and Sofic Groups

In [131] Rădulescu considered the particular case of Connes' embedding conjecture for II_1 factors arising as group factors $L\Gamma$ of countable discrete ICC groups. A countable discrete ICC group Γ is called *hyperlinear* if $L\Gamma$ verifies Connes' embedding conjecture. As we will see in Sects. 2.2 and 2.6 the notion of hyperlinear group admits several equivalent characterizations, as well as a natural generalization to the class of all countable groups. The notion of hyperlinear group turned out to be tightly connected with the notion of *sofic groups*. Sofic groups are a class of countable discrete groups introduced by Gromov in [79] (even though the name "sofic" was coined by Weiss in [150]). A group is sofic if, loosely speaking, it can be locally approximated by finite permutation groups S_n up to an error measured in terms of the Hamming metric S_n (see Sect. 2.1). Elek and Szabó in [54] showed that every sofic group is hyperlinear. Gromov's motivation to introduce the notion of sofic groups came from an open problem in symbolic dynamics known as Gottschalk's surjunctivity conjecture: Suppose that Γ is a countable group and A is a finite set. Denote by A^Γ the set of Γ-sequences of elements of A. The product topology on A^Γ with respect to the discrete topology on A is compact and metrizable.

The *Bernoulli shift* of Γ with alphabet A is the left action of Γ on A^Γ defined by

$$\rho \cdot \left(a_\gamma\right)_{\gamma \in \Gamma} = \left(a_{\rho^{-1}\gamma}\right)_{\gamma \in \Gamma}.$$

A continuous function $f : A^\Gamma \to A^\Gamma$ is *equivariant* if it preserves the Bernoulli action, i.e. $f(\rho \cdot x) = \rho \cdot f(x)$, for every $x \in A^\Gamma$ and $\rho \in \Gamma$. Conjecture 1.4.1 was proposed by Gottschalk in [76] and it is usually referred to as Gottschalk's surjunctivity conjecture.

Conjecture 1.4.1 Suppose that Γ is a discrete group and A is a finite set. If $f : A^\Gamma \to A^\Gamma$ is a continuous injective equivariant function, then f is surjective.

It is currently an open problem to determine whether Gottschalk's surjunctivity conjecture holds for all countable discrete groups. Gromov proved in [79] using graph-theoretical methods that sofic groups satisfy Gottschalk's surjunctivity conjecture. Another proof was obtained Kerr and Li in [99] as an application of the theory of entropy for actions of sofic groups developed by Bowen, Kerr, and Li (see [15, 99, 100]). The countable discrete groups satisfying Gottschalk's surjunctivity conjecture are sometimes called *surjunctive*. An example of a *monoid* that does not satisfy the natural generalization of surjunctivity for monoids has been provided in [33]. More information about Gottschalk's surjunctivity conjecture and surjunctive groups can be found in the monograph [31].

Since Gromov's proof of Gottschalk's conjecture for sofic groups, many other open problems have been settled for sofic or hyperlinear groups such as the Kervaire-Laudenbach conjecture (see Sect. 2.8), Kaplansky's direct finiteness conjecture (see Sects. 1.5 and 2.10), and the algebraic eigenvalues conjecture (see Sect. 2.12). This showed how deep and fruitful the notion of hyperlinear and sofic groups are, and contributed to bring considerable attention to the following question which is strikingly still open:

Open Problem 1.4.2 Is there any countable discrete group Γ that is not sofic or hyperlinear?

1.5   Kaplansky's Direct Finiteness Conjecture

Suppose that Γ is a countable discrete group, and consider the complex group algebra $\mathbb{C}\Gamma$ defined in Sect. 1.1. Kaplanksi showed in [93] that $\mathbb{C}\Gamma$ is a directly finite ring. This means that if $a, b \in \mathbb{C}\Gamma$ are such that $ab = 1$ then $ba = 1$. In [19] Burger and Valette gave a short proof of Kaplansky's result by means of the group von Neumann algebra construction introduced in Sect. 1.1. Indeed, one can regard $\mathbb{C}\Gamma$ as a subalgebra of the group von Neumann algebra $L\Gamma$ and prove that $L\Gamma$ is directly finite using analytic methods. This is the content of Theorem 1.5.1.

Theorem 1.5.1 *If M is a von Neumann algebra endowed with a faithful finite trace τ, then M is a directly finite algebra.*

We now prove this theorem. Let M be a von Neumann algebra and τ be a faithful normalized trace on M. If $x, y \in M$ are such that $xy = 1$ then $yx \in M$ is an idempotent element such that

$$\tau(yx) = \tau(xy) = \tau(1) = 1.$$

It is thus enough to prove that if $e \in M$ is an idempotent element such that $\tau(e) = 1$ then $e = 1$. This is equivalent to the assertion that if $e \in M$ is an idempotent element such that $\tau(e) = 0$ then $e = 0$. The latter statement is proved in Lemma 1.5.2 (see [19, Lemma 2.1]).

Lemma 1.5.2 *If M is a von Neumann algebra endowed with a faithful finite trace τ and $e \in M$ is an idempotent such that $\tau(e) = 0$, then $e = 0$.*

Proof The conclusion is obvious if e is a self-adjoint idempotent element (i.e. a projection). In fact in this case

$$\tau(e) = \tau\left(e^*e\right) = 0$$

implies $e = 0$ by faithfulness of τ. In order to establish the general case it is enough to show that if $e \in M$ is idempotent, then there is a self-adjoint invertible element z of M such that $f = ee^*z^{-1}$ is a projection and $\tau(e) = \tau(f)$. Define

$$z = 1 + \left(e^* - e\right)^* \left(e^* - e\right).$$

Observe that z is an invertible element (see [14, II.3.1.4]) commuting with e. It is not difficult to check that $f = ee^*z^{-1}$ has the required properties. □

The construction of the complex group algebra of Γ introduced in Sect. 1.1 can be generalized replacing $\mathbb{C}$ with an arbitrary field K. One thus obtains the group K-algebra $K\Gamma$ of Γ consisting of formal finite linear combinations

$$k_1\gamma_1 + \ldots + k_n\gamma_n$$

of elements of Γ with coefficients in K. As before a typical element a of $K\Gamma$ can be denoted by

$$\sum_{\gamma} a_\gamma \gamma.$$

where the coefficients $a_\gamma \in K$ are zero for all but finitely many $\gamma \in \Gamma$. The operations on $K\Gamma$ are defined by

$$\left(\sum_\gamma a_\gamma \gamma\right) + \left(\sum_\gamma b_\gamma \gamma\right) = \sum_\gamma (a_\gamma + b_\gamma)\,\gamma$$

and

$$\left(\sum_\gamma a_\gamma \gamma\right)\left(\sum_\gamma b_\gamma \gamma\right) = \sum_\gamma \left(\sum_{\rho\rho'=\gamma} a_\rho b_{\rho'}\right)\gamma.$$

Conjecture 1.5.3 is due to Kaplansky and usually referred to as direct finiteness conjecture.

Conjecture 1.5.3 The algebra $K\Gamma$ is directly finite, i.e. any one-side invertible element of $K\Gamma$ is invertible.

Kaplansky's direct finiteness conjecture was established for residually amenable groups in [3] and then for sofic groups in [53]. Section 2.10 contains a proof of Conjecture 1.5.3 for sofic groups involving the notion of rank ring and rank ultraproduct of rank rings. It is a well know fact (see Observation 2.1 in [126]) that any field embeds as a subfield of an ultraproduct of finite fields (see Sect. 1.2 of for the definition of ultraproduct of fields). It is not difficult to deduce from this observation that a group Γ satisfies Kaplansky's direct finiteness conjecture as soon as $K\Gamma$ is directly finite for every finite field K (this fact was pointed out to us by Vladimir Pestov). This allows one to deduce that Gottschalk's conjecture is stronger than Kaplansky's direct finiteness conjecture. In fact suppose that Γ is a group satisfying Gottschalk's conjecture and K is a field that, without loss of generality, we can assume to be finite. Consider the Bernoulli action of Γ with alphabet K. We will denote an element $(a)_{\gamma\in\Gamma}$ of K^Γ by $\sum_\gamma a_\gamma \gamma$, and regard the group algebra $K\Gamma$ as a subset of K^Γ. Defining

$$\left(\sum_\gamma a_\gamma \gamma\right) \cdot \left(\sum_\gamma b_\gamma \gamma\right) = \sum_\gamma \left(\sum_{\rho\rho'=\gamma} a_\rho b_{\rho'}\right)\gamma$$

for $\sum_\gamma a_\gamma \gamma \in K^\Gamma$ and $\sum_\gamma b_\gamma \gamma \in K\Gamma$ one obtains a right action of $K\Gamma$ on K^Γ that extends the multiplication operation in $K\Gamma$ and commutes with the left action of Γ on $K\Gamma$. Suppose that $a, b \in K\Gamma$ are such that $ab = 1_\Gamma$. Define the continuous equivariant map $f : K^\Gamma \to K^\Gamma$ by $f(x) = x \cdot a$. It follows from the fact that b is a right inverse of a that f is injective. Since Γ is assumed to satisfy Gottschalk's conjecture, f must be surjective. In particular there is $x_0 \in K^\Gamma$ such that

$$x_0 \cdot a = f(x_0) = 1_\Gamma.$$

It follows that a has a left inverse. A standard calculation allows one to conclude that a is invertible with inverse b.

1.6 Kaplansky's Group Ring Conjectures

Conjecture 1.5.3 is only one of several conjectures concerning the group algebra $K\Gamma$ for a countable discrete *torsion-free* group Γ attributed to Kaplansky. Here is the complete list:

- Zero divisors conjecture: $K\Gamma$ has no zero divisors;
- Nilpotent elements conjecture: $K\Gamma$ has no nilpotent elements;
- Idempotent elements conjecture: the only idempotent elements of $K\Gamma$ are 0 and 1;
- Units conjecture: the only units of $K\Gamma$ are $k\gamma$ for $k \in K \setminus \{0\}$ and $\gamma \in \Gamma$;
- Trace of idempotents conjecture: if b is an idempotent element of $K\Gamma$ then the coefficient b_{1_Γ} corresponding to the identity 1_Γ of Γ belongs to the prime field K_0 of K (which is the minimum subfield of K).

All these conjectures are to this day still open, apart from the last one which has been established by Zaleesskii in [151]. We will present now a proof of the particular case of Zaleesskii's result when K is a finite field of characteristic p. Recall that a *trace* on $K\Gamma$ is a K-linear map $\tau : K\Gamma \to K$ such that $\tau(ab) = \tau(ba)$ for every $a, b \in K\Gamma$. If n is a nonnegative integer and $a \in K\Gamma$, define $\tau_n(a)$ to be the sum of the coefficients of a corresponding to elements of Γ of order p^n. In particular $\tau_0(a)$ is the coefficient of a corresponding t the identity element 1_Γ of Γ.

Exercise 1.6.1 Verify that τ_n is a trace on $K\Gamma$.

Exercise 1.6.2 Show that if K is a finite field of characteristic p and τ is any trace on $K\Gamma$, then

$$\tau\left((a+b)^p\right) = \tau(a^p) + \tau(b^p), \tag{1.2}$$

for every $a, b \in K\Gamma$. Thus by induction

$$\tau(a^p) = \sum_\gamma a_\gamma^p \tau(\gamma^p).$$

The identity (1.2) can be referred to as "Frobenius under trace" in analogy with the corresponding identity for elements of a field of characteristic p. Suppose now that $e \in K\Gamma$ is an idempotent element. We want to show that $\tau(e)$ belongs to the prime field K_0 of K. To this purpose it is enough to show that $\tau(e)^p = \tau(e)$. For

$n \geq 1$ we have, by Exercises 1.6.1 and 1.6.2, denoting by $|\gamma|$ the order of an element γ of Γ:

$$
\begin{aligned}
\tau_n(e) &= \tau_n(e^p) \\
&= \sum_\gamma e_\gamma^p \tau_n(\gamma^p) \\
&= \sum_{|\gamma|=p^{n+1}} e_\gamma^p \\
&= \left(\sum_{|\gamma|=p^{n+1}} e_\gamma \right)^p \\
&= \tau_{n+1}(e)^p.
\end{aligned}
$$

On the other hand:

$$
\begin{aligned}
\tau_0(e) &= \tau_0(e^p) \\
&= \sum_\gamma e_\gamma^p \tau_0(\gamma^p) \\
&= \sum_{|\gamma|=1} e_\gamma^p + \sum_{|\gamma|=p} e_\gamma^p \\
&= \tau_0(e)^p + \tau_1(e)^p.
\end{aligned}
$$

From these identities it is easy to prove by induction that

$$
\tau_0(e) = \tau_0(e)^p + \tau_n(e)^{p^n},
$$

for every $n \in \mathbb{N}$. Since e has finite support, there is $n \in \mathbb{N}$ such that $\tau_n(e) = 0$. This implies that $\tau_0(e) = \tau_0(e)^p$ and hence $\tau_0(e) \in K_0$.

The proof of the general case of Zalesskii's theorem can be inferred from the particular case presented here. The details can be found in [19].

Chapter 2
Sofic and Hyperlinear Groups

Martino Lupini

2.1 Definition of Sofic Groups

A *length function* ℓ on a group G is a function $\ell : G \to [0, 1]$ such that for every $x, y \in G$:

- $\ell(xy) \leq \ell(x) + \ell(y)$;
- $\ell(x^{-1}) = \ell(x)$;
- $\ell(x) = 0$ if and only if x is the identity 1_G of G.

A length function is called *invariant* if it is moreover invariant by conjugation. This means that for every $x, y \in G$

$$\ell(xyx^{-1}) = \ell(x)$$

or equivalently

$$\ell(xy) = \ell(yx).$$

A group endowed with an invariant length function is called an *invariant length group*. If G is an invariant length group with invariant length function ℓ, then the function

$$d : G \times G \to [0, 1]$$

M. Lupini (✉)
Department of Mathematics, California Institute of Technology, 1200 East California Boulevard, Pasadena, CA 91125, USA
e-mail: mlupini@mathstat.yorku.ca

© Springer International Publishing Switzerland 2015
V. Capraro, M. Lupini, *Introduction to Sofic and Hyperlinear Groups and Connes' Embedding Conjecture*, Lecture Notes in Mathematics 2136, DOI 10.1007/978-3-319-19333-5_2

defined by $d(x, y) = \ell(xy^{-1})$ is a bi-invariant metric on G. This means that d is a metric on G such that left and right translations in G are isometries with respect to d. Conversely any bi-invariant metric d on G gives rise to an invariant length function ℓ on G by

$$\ell(x) = d(x, 1_G).$$

This shows that there is a bijective correspondence between invariant length functions and bi-invariant metrics on a group G. If Γ is any group, then the function

$$\ell_0(x) = \begin{cases} 0 & \text{if } x = 1_G, \\ 1 & \text{otherwise;} \end{cases}$$

is an invariant length function on Γ, called the trivial invariant length function. In the following any discrete group will be regarded as an invariant length group endowed with the trivial invariant length function.

Consider for $n \in \mathbb{N}$ the group S_n of permutations over the set $n = \{0, 1, \ldots, n-1\}$. The *Hamming invariant length function* ℓ on S_n is defined by

$$\ell_{S_n}(\sigma) = \frac{1}{n} |\{i \in n : \sigma(i) \neq i\}|.$$

Exercise 2.1.1 Verify that ℓ_{S_n} is in fact an invariant length function on S_n.

The bi-invariant metric on S_n associated with the invariant length function ℓ_{S_n} will be denoted by d_{S_n}.

Definition 2.1.2 A countable discrete group Γ is *sofic* if for every $\varepsilon > 0$ and every finite subset F of $\Gamma \setminus \{1_\Gamma\}$ there is a natural number n and a function $\Phi : \Gamma \to S_n$ such that $\Phi(1_\Gamma) = 1_{S_n}$ and for every $g, h \in F \setminus \{1_\Gamma\}$:

- $d_{S_n}(\Phi(gh), \Phi(g)\Phi(h)) < \varepsilon$;
- $\ell_{S_n}(\Phi(g)) > r(g)$ where $r(g)$ is a positive constant depending only on g.

This *local approximation property* can be reformulated in terms of embedding into a (length) ultraproduct of the permutation groups. The product

$$\prod_{n \in \mathbb{N}} S_n$$

is a group with respect to the coordinatewise multiplication. Fix a free ultrafilter $\mathcal{U}$ over $\mathbb{N}$ (see Appendix B for an introduction to ultrafilters). Define the function

$$\ell_{\mathcal{U}} : \prod_{n \in \mathbb{N}} S_n \to [0, 1]$$

by

$$\ell_{\mathcal{U}}((\sigma_n)_{n \in \mathbb{N}}) = \lim_{n \to \mathcal{U}} \ell_{S_n}(\sigma_n).$$

It is not hard to check

$$N_{\mathcal{U}} = \left\{ x \in \prod_{n \in \mathbb{N}} S_n : \ell_{\mathcal{U}}(x) = 0 \right\}$$

is a normal subgroup of $\prod_n S_n$. The quotient of $\prod_n S_n$ by $N_{\mathcal{U}}$ is denoted by

$$\prod_{\mathcal{U}} S_n$$

and called the *ultraproduct* relative to the free ultrafilter $\mathcal{U}$ of the sequence $(S_n)_{n \in \mathbb{N}}$ of permutation groups regarded as invariant length groups.

Exercise 2.1.3 Show that if $(\sigma_n)_{n \in \mathbb{N}}, (\tau_n)_{n \in \mathbb{N}} \in \prod_n S_n$ belong to the same coset of $N_{\mathcal{U}}$ then $\ell_{\mathcal{U}}((\sigma_n)_{n \in \mathbb{N}}) = \ell_{\mathcal{U}}((\tau_n)_{n \in \mathbb{N}})$.

Exercise 2.1.3 shows that the function $\ell_{\mathcal{U}}$ passes to the quotient inducing a canonical invariant length function on $\prod_{\mathcal{U}} S_n$ still denoted by $\ell_{\mathcal{U}}$. If $x \in \prod_{\mathcal{U}} S_n$ belongs to the coset of $N_{\mathcal{U}}$ associated with $(\sigma_n)_{n \in \mathbb{N}} \in \prod_n S_n$, then $(\sigma_n)_{n \in \mathbb{N}}$ is called a *representative sequence* for the element x. It is not difficult to reformulate the notion of sofic group in term of existence of an embedding into $\prod_{\mathcal{U}} S_n$.

Exercise 2.1.4 Suppose that Γ is a countable discrete group. Show that the following statements are equivalent:

1. Γ is sofic;
2. there is an injective group *-homomorphism $\Phi : \Gamma \to \prod_{\mathcal{U}} S_n$ for every free ultrafilter $\mathcal{U}$ over N;
3. there is an injective group *-homomorphism $\Phi : \Gamma \to \prod_{\mathcal{U}} S_n$ for some free ultrafilter $\mathcal{U}$ over N.

Hint For 1. $\Rightarrow$ 2. observe that the hypothesis implies that there is a sequence $(\Phi_n)_{n \in \mathbb{N}}$ of maps from Γ to S_n such that $\Phi_n(1_\Gamma) = 1_{S_n}$ and for every $g, h \in \Gamma \setminus \{1_\Gamma\}$

$$\lim_{n \to +\infty} d_{S_n}(\Phi_n(gh), \Phi_n(h)\Phi_n(g)) = 0$$

and

$$\liminf_{n \to +\infty} \ell_{S_n}(\Phi_n(g)) \geq r(g) > 0.$$

Define $\Phi : \Gamma \to \prod_{\mathcal{U}} S_n$ sending g to the element of $\prod_{\mathcal{U}} S_n$ having $(\Phi_n(g))_{n \in \mathbb{N}}$ as representative sequence. For 3. $\Rightarrow$ 1. observe that if $\Phi : \Gamma \to \prod_{\mathcal{U}} S_n$ is an injective *-homomorphism and for every $g \in G$

$$(\Phi_n(g))_{n \in \mathbb{N}}$$

is a representative sequence of $\Phi(g)$ then the maps $\Psi_n = \Phi_n(1_\Gamma)^{-1} \Phi_n(g)$ satisfy the following properties: $\Psi_n(1_\Gamma) = 1_{S_n}$ and for every $g, h \in \Gamma \setminus \{1_\Gamma\}$

$$\lim_{n \to \mathcal{U}} d_{S_n}(\Psi_n(gh), \Psi_n(g)\Psi_n(h)) = 0$$

and

$$\lim_{n \to \mathcal{U}} \ell_{S_n} (\Psi_n (g)) \geq \ell_{\mathcal{U}} (\Phi (g)).$$

If $F \subset \Gamma$ is finite and $\varepsilon > 0$ then the maps Ψ_n for n large enough witness the condition of soficity of Γ relative to F and $\varepsilon > 0$. ∎

In view of the fact that a (countable, discrete) group is sofic if and only embeds in $\prod_{\mathcal{U}} S_n$ for some free ultrafilter $\mathcal{U}$, the (length) ultraproducts of the sequence of permutations are called *universal sofic groups* (see [54, 111, 124, 126, 145]).

Universal sofic groups are not separable. This fact follows from a well known argument. Two functions $f, g : \mathbb{N} \to \mathbb{N}$ are *eventually distinct* if they coincide only on finitely many n's.

Exercise 2.1.5 Suppose that $h : \mathbb{N} \to \mathbb{N}$ is an unbounded function. Show that there is a family $\mathcal{F}$ of size continuum of pairwise eventually distinct functions from $\mathbb{N}$ to $\mathbb{N}$ such that $f(n) \leq h(n)$ for every $f \in \mathcal{F}$.

Hint For every subset A of $\mathbb{N}$ define $f_A : \mathbb{N} \to \mathbb{N}$ by

$$f_A (n) = \sum_{k < n} \chi_A (n) 2^k$$

where χ_A denotes the characteristic function of A. Observe that

$$\mathcal{F}_0 = \{f_A : A \subset \mathbb{N}\}$$

is a family of size continuum of pairwise eventually distinct functions such that $f(n) \leq 2^n$ for every $n \in \mathbb{N}$. ∎

Exercise 2.1.6 Show that for any free ultrafilter $\mathcal{U}$ there is a subset X of $\prod_{\mathcal{U}} S_n$ of size continuum such that every two distinct elements of X have distance one. Conclude that any dense subset of $\prod_{\mathcal{U}} S_n$ has the cardinality of the continuum.

Hint By Exercise 2.1.5 there is a family $\mathcal{F}$ of size continuum of pairwise eventually distinct functions such that $f(n) \leq \sqrt{n}$ for every $n \in \mathbb{N}$ and $f \in \mathcal{F}$. For every $f \in \mathcal{F}$ consider the element x_f of $\prod_{\mathcal{U}} S_n$ having

$$\left(\sigma_n^{f(n)} \right)_{n \in \mathbb{N}}$$

as representative sequence, where σ_n is any cyclic permutation of n. Show that

$$X := \left\{ x_f : f \in \mathcal{F} \right\}$$

has the required property. ∎

An amplification argument of Elek and Szabó (see [54]) shows that the condition of soficity is equivalent to the an apparently stronger approximation property, which is discussed in the following Exercise 2.1.8.

Definition 2.1.7 Suppose that G, H are invariant length groups, F is a subset of G, and ε is a positive real number. A function $\Phi : G \to H$ is called an (F, ε)-*approximate morphism* if $\Phi(1_G) = 1_H$ and for every $g, h \in F$:

- $\left| \ell_H \left(\Phi(gh)\Phi(h)^{-1}\Phi(g) \right) \right| < \varepsilon$;
- $\left| \ell_H \left(\Phi(g) \right) - \ell_G(g) \right| < \varepsilon$.

Exercise 2.1.8 Prove that a countable discrete group Γ is sofic if and only if for every positive real number ε and every finite subset F of Γ there is a natural number n and an (F, ε)-approximate morphism from Γ to S_n, where Γ is regarded as an invariant length group with respect to the trivial invariant length function.

Hint If $n, k \in \mathbb{N}$ and $\sigma \in S_n$ consider the permutation $\sigma^{\otimes k}$ of the set $[n]^k$ of k-sequences of elements of n defined by

$$\sigma^{\otimes k}(i_1, \ldots, i_k) = (\sigma(i_1), \ldots, \sigma(i_k)).$$

Identifying the group of permutations of $\overbrace{n \times \cdots \times n}^{k \text{ times}}$ with S_{n^k}, the function

$$\sigma \mapsto \sigma^{\otimes k}$$

defines a group *-homomorphism from S_n to S_{n^k} such that

$$1 - \ell_{S_{n^k}}\left(\sigma^{\otimes k}\right) = \left(1 - \ell_{S_n}(\sigma)\right)^k.$$

∎

Using Exercise 2.1.8 one can express the notion of soficity in terms of length-preserving embedding into ultraproducts of permutations groups.

Exercise 2.1.9 Suppose that Γ is a countable discrete group regarded as an invariant length group endowed with the trivial invariant length. Show that the following statements are equivalent:

- Γ is sofic;
- there is a length-preserving *-homomorphism $\Phi : \Gamma \to \prod_{\mathcal{U}} S_n$ for every free ultrafilter $\mathcal{U}$ over N;
- there is an length-preserving *-homomorphism $\Phi : \Gamma \to \prod_{\mathcal{U}} S_n$ for some free ultrafilter $\mathcal{U}$ over N.

Hint Follow the same steps as in the proof of Exercise 2.1.9, replacing the condition given in the definition of sofic group with the equivalent condition expressed in Exercise 2.1.8. ∎

In a number of cases it is useful to consider approximate morphisms satisfying convenient additional properties. An example of such morphisms with additional properties is considered in the following exercise, originally proved in [55, Lemma 2.1].

Exercise 2.1.10 Suppose that Γ is a countable discrete sofic group, F is a finite symmetric subset of Γ containing the identity element, and ε is a positive real number. Show that one can find an (F, ε)-approximate morphism $g \mapsto \sigma_g$ from Γ to S_n for some $n \in \mathbb{N}$ such that, for every $g \in F$, σ_g has no fixed points and $\sigma_{g^{-1}} = \sigma_g^{-1}$.

Hint Fix $\eta > 0$ small enough and set $\hat{F} = F \cdot F$. By Exercise 2.1.8 one can consider an $(\hat{F}, \eta)$-approximate morphism $g \mapsto \tau_g$ from Γ to S_n for some $n \in \mathbb{N}$. For any nonidentity element g of $\hat{F}$ consider the largest subset A_g of n such that $\tau_{g^{-1}}\tau_g$ is the identity on A_g and τ_g has no fixed points on A_g. Verify that τ_g defines a bijection between A_g and $A_{g^{-1}}$. Define an approximate morphism $g \mapsto \sigma_g$ from Γ to S_{2n} in the following way. Identify S_{2n} with the group of premutations of $n \times 2$. For any nonidentity element g of $\hat{F}$ fix a bijection ϕ_g from $A_g \backslash A_{g^{-1}} \times 2$ to $A_{g^{-1}} \backslash A_g \times 2$ and a fixed-point free involution ψ_g of $\left(n \backslash (A_g \cup A_{g^{-1}}) \right) \times 2$ such that $\phi_{g^{-1}} = \phi_g^{-1}$ and $\psi_{g^{-1}} = \psi_g^{-1}$. Define then

$$
\tau_g(i,j) = \begin{cases} \left(\sigma_g(i), j \right) & \text{if } i \in A_g, \\ \phi_g(i,j) & \text{if } i \in A_g \backslash A_{g^{-1}}, \\ \phi_{g^{-1}}(i,j) & \text{if } i \in A_{g^{-1}} \backslash A_g, \\ \psi_g(i,j) & \text{otherwise.} \end{cases}
$$

Verify that for η small enough the assignment $g \mapsto \tau_g$ is an (F, ε)-approximate morphism with the required extra properties. ∎

2.2 Definition of Hyperlinear Groups

Recall that $M_n(\mathbb{C})$ denotes the tracial von Neumann algebra of $n \times n$ matrices over the complex numbers. The *normalized trace* τ of $M_n(\mathbb{C})$ is defined by

$$
\tau \left((a_{ij}) \right) = \frac{1}{n} \sum_{i=1}^{n} a_{ii}.
$$

The *Hilbert-Schmidt norm* $\|x\|_2$ on $M_n(\mathbb{C})$ is defined by

$$
\|x\|_2 = \tau(x^*x)^{\frac{1}{2}}.
$$

An element x of $M_n(\mathbb{C})$ is unitary if $x^*x = xx^* = 1$. The set U_n of unitary elements of $M_n(\mathbb{C})$ is a group with respect to multiplication. The *Hilbert-Schmidt invariant length function* on U_n is defined by

$$
\ell_{U_n}(u) = \frac{1}{2} \|u - 1\|_2.
$$

Exercise 2.2.1 Show that ℓ_{U_n} is an invariant length function on U_n such that $\ell_{U_n}(u)^2 = \frac{1}{2}(1 - \operatorname{Re}(\tau(u)))$.

Hyperlinear groups are defined exactly as sofic groups, where the permutation groups with the Hamming invariant length function are replaced by the unitary groups endowed with the Hilbert-Schmidt invariant length function.

Definition 2.2.2 A countable discrete group Γ is *hyperlinear* if for every $\varepsilon > 0$ and every finite subset F of $\Gamma \setminus \{1_\Gamma\}$ there is a natural number n and a function $\Phi : \Gamma \to U_n$ such that $\Phi(1_\Gamma) = 1_{U_n}$ and for every $g, h \in F$:

- $d_{U_n}(\Phi(gh), \Phi(g)\Phi(h)) < \varepsilon$;
- $\ell_{U_n}(\Phi(g)) > r(g)$ where $r(g)$ is some positive constant depending only on g.

As before this notion can be equivalently reformulated in terms of embedding into ultraproducts. If $\mathcal{U}$ is a free ultrafilter over $\mathbb{N}$ the ultraproduct $\prod_{\mathcal{U}} U_n$ of the unitary groups regarded as invariant length groups is the quotient of $\prod_n U_n$ with respect to the normal subgroup

$$N_{\mathcal{U}} = \left\{ (u_n)_{n \in \mathbb{N}} : \lim_{n \to \mathcal{U}} \ell_{U_n}(u_n) = 0 \right\}$$

endowed with the invariant length function

$$\ell_{\mathcal{U}}((u_n) N_{\mathcal{U}}) = \lim_{n \to \mathcal{U}} \ell(u_n).$$

Exercise 2.2.3 Suppose that Γ is a countable discrete group. Show that the following statements are equivalent:

- Γ is hyperlinear;
- there is an injective *-homomorphism $\Phi : \Gamma \to \prod_{\mathcal{U}} U_n$ for every free ultrafilter $\mathcal{U}$ over N;
- there is an injective *-homomorphism $\Phi : \Gamma \to \prod_{\mathcal{U}} U_n$ for some free ultrafilter $\mathcal{U}$ over N.

As in the case of sofic groups, (length) ultraproducts of sequences of unitary groups can be referred to as *universal hyperlinear groups* in view of Exercise 2.2.3. Exercise 3.2.1 in Sect. 3.2 shows that the same conclusion of Exercise 2.2.3 holds for (length) ultrapowers of the unitary group of the hyperfinite II_1 factor (see Definition 3.1.2).

If σ is a permutation over n denote by P_σ the permutation matrix associated with σ, acting as σ on the canonical basis of $\mathbb{C}^n$. Observe that P_σ is a unitary matrix and the function

$$\sigma \mapsto P_\sigma$$

is a *-homomorphism from S_n to U_n. Moreover

$$\tau(P_\sigma) = 1 - \ell_{S_n}(\sigma)$$

and hence

$$\ell_{U_n}(P_\sigma)^2 = \frac{1}{2}\ell_{S_n}(\sigma). \tag{2.1}$$

It is not difficult to deduce from this that any sofic group is hyperlinear. This is the content of Exercise 2.2.4.

Exercise 2.2.4 Fix a free ultrafilter $\mathcal{U}$ over N. Show that the function

$$(\sigma_n)_{n\in\mathbb{N}} \mapsto (P_{\sigma_n})_{n\in\mathbb{N}}$$

from $\prod_n S_n$ to $\prod_n U_n$ induces an algebraic embedding of $\prod_{\mathcal{U}} S_n$ into $\prod_{\mathcal{U}} U_n$.

Proposition is an immediate consequence of Exercise 2.2.4, together with Exercises 2.1.4 and 2.2.3.

Proposition 2.2.5 *Every countable discrete sofic group is hyperlinear.*

It is currently and open problem to determine whether the notions of sofic and hyperlinear group are actually distinct, since no example of a nonsofic groups is known.

Exercise 2.2.6 Show that for any free ultrafilter $\mathcal{U}$ there is a subset X of $\prod_{\mathcal{U}} U_n$ of size continuum such that every pair of distinct elements of X has distance $\frac{1}{\sqrt{2}}$.

Hint Proceed as in Exercise 2.1.6, where for every $n \in \mathbb{N}$ the cyclic permutation σ_n is replaced with the unitary permutation matrix P_{σ_n} associated with σ_n. Recall the relation between the Hamming length function and the Hilbert-Schmidt length functions given by Eq. (2.1). ∎

It follows from Exercise 2.2.6 that a dense subset of $\prod_{\mathcal{U}} U_n$ has the cardinality of the continuum. In particular the universal hyperlinear groups are not separable.

An amplification argument from [131] due to Rădulescu (predating the analogous argument for permutation groups of Elek and Szabó) shows that hyperlinearity is equivalent to an apparently stronger approximation property. In order to present this argument we need to recall some facts about tensor product of matrix algebras.

If $M_n(\mathbb{C})$ and $M_m(\mathbb{C})$ are the algebras of, respectively, $n \times n$ and $m \times m$ matrices with complex coefficients, then their tensor product (see Appendix A) can be canonically identified with the algebra $M_{nm}(\mathbb{C})$ of $nm \times nm$ matrices. Concretely if $A = (a_{ij}) \in M_n(\mathbb{C})$ and $B = (b_{ij}) \in M_m(\mathbb{C})$ then $A \otimes B$ is identified with the block matrix

$$\begin{pmatrix} a_{11}B & a_{12}B & \dots & a_{1n}B \\ a_{21}B & a_{22}B & \dots & a_{2n}B \\ \vdots & \vdots & \ddots & \vdots \\ a_{n1}B & \dots & \dots & a_{nn}B \end{pmatrix}.$$

It is easily verified that

$$\tau (A \otimes B) = \tau(A)\tau (B)$$

and $A \otimes B$ is unitary if both A and B are.

Exercise 2.2.7 Suppose that $u \in U_n$. If u is different from the identity, then the absolute value of the trace of

$$\begin{pmatrix} u & 0 \\ 0 & 1_{U_n} \end{pmatrix} \in U_{2n}$$

is strictly smaller than 1.

Exercise 2.2.8 If $u \in U_n$ then define recursively $u^{\otimes 1} = u \in U_n$ and $u^{\otimes (k+1)} = u^{\otimes k} \otimes u \in U_{n^{k+1}}$. Show that the function

$$u \mapsto u^{\otimes k}$$

is a group *-homomorphism from U_n to U_{n^k} such that

$$\tau(u) = \tau(u)^k.$$

We can now state and prove the promised equivalent characterization of hyperlinear groups.

Proposition 2.2.9 *A countable discrete group Γ is hyperlinear if and only for every finite subset F of Γ and every positive real number ε there is a natural number n and an (F, ε)-approximate morphism from Γ to U_n, where Γ is regarded as invariant length groups with respect to the trivial length function.*

Proof Suppose that Γ is hyperlinear. If F is a finite subset of Γ and ε is a positive real number, consider the map $\Phi : \Gamma \to U_n$ obtained from F and ε as in the definition of hyperlinear group. By Exercise 2.2.7 after replacing Φ with the map

$$\Gamma \to U_{2n}$$

$$\gamma \mapsto \begin{pmatrix} \Phi(\gamma) & 0 \\ 0 & 1_{U_n} \end{pmatrix}$$

we can assume that

$$|\tau (\Phi(\gamma))| < 1$$

for every $\gamma \in F$. It is now easy to show using Exercise 2.2.8 that the map

$$\Gamma \to U_{n^k}$$

$$\gamma \mapsto \Phi(\gamma)^{\otimes k}$$

for $k \in \mathbb{N}$ large enough satisfies the requirements of the statement. The converse implication is obvious.   □

As before we can deduce a characterization of hyperlinear groups in terms of length-preserving embeddings into ultraproducts of unitary groups.

Exercise 2.2.10 Suppose that Γ is a countable discrete group regarded as an invariant length group endowed with the trivial invariant length function. Show that the following statements are equivalent:

- Γ is hyperlinear;
- there is a length preserving *-homomorphism $\Phi : \Gamma \to \prod_{\mathcal{U}} U_n$ for every free ultrafilter $\mathcal{U}$ over N;
- there is a length preserving *-homomorphism $\Phi : \Gamma \to \prod_{\mathcal{U}} U_n$ for some free ultrafilter $\mathcal{U}$ over N.

We conclude this section with yet another reformulation of the definition of hyperlinear group in terms of what are usually called *microstates*. Suppose that Γ is a finitely generated group with finite generating set $S = \{g_1, \ldots, g_m\}$. If $n \in \mathbb{N}$ and $\varepsilon > 0$ then an (n, ε)-microstate for Γ is a map $\Phi : \Gamma \to M_k(\mathbb{C})$ for some $k \in \mathbb{N}$ such that for every word w in x_i and x_i^{-1} for $i \leq m$ of length at most n one has that

$$|\tau (w (\Phi (g_1), \ldots, \Phi (g_m)))| < \varepsilon \quad \text{whenever } w (g_1, \ldots, g_m) \neq 1_\Gamma,$$

and

$$|\tau (w (\Phi (g_1), \ldots, \Phi (g_m))) - 1| < \varepsilon \text{ whenever } w (g_1, \ldots, g_n) = 1_\Gamma.$$

A countable discrete groups is hyperlinear if and only if it admits (n, ε)-microstates for every $n \in \mathbb{N}$ and $\varepsilon > 0$. In order to verify that such a definition is equivalent to the previous ones—and, in particular, does not depend on the choice of the generating set S—let us consider the *relation $R(x)$* defined by

$$\max \{\|xx^* - 1\|_2, \|x^*x - 1\|_2\}.$$

It is obvious that an element a in $M_n(\mathbb{C})$ is a unitary if and only if $R(a) = 0$. Moreover such a relation has the following property, usually called *stability*: for every $\varepsilon > 0$ there is $\delta > 0$ such that for every $n \in \mathbb{N}$ and every element a of $M_n(\mathbb{C})$ if $R(a) < \delta$ then there is a unitary $u \in M_n(\mathbb{C})$ such that $\|a - u\|_2 < \varepsilon$. In other words any approximate solution to the equation $R(x) = 0$ is close to an exact solution. Such a property, which holds even if one replaces $M_n(\mathbb{C})$ with an arbitrary tracial von Neumann algebra, can be established by means of the *polar decomposition* of an element inside a von Neumann algebra; see [14, Sect. III.1.1.2].

Exercise 2.2.11 Using stability of the relation defining unitary elements, verify that the microstates formulation of hyperlinearity is equivalent to the original definition.

2.3 Classes of Sofic and Hyperlinear Groups

A classical theorem of Cayley (see [29]) asserts that any finite group is isomorphic to a group of permutations (with no fixed points) on a finite set. To see this just let the group act on itself by left translation. This observation implies in particular that finite groups are sofic. This argument can be generalized to prove that amenable groups are sofic. Recall that a countable discrete group Γ is *amenable* if for every finite subset F of Γ and for every $\varepsilon > 0$ there is an (H, ε)-invariant finite subset K of Γ, i.e. such that

$$|hK \triangle K| < \varepsilon |K|$$

for every $h \in H$. Amenable groups were introduced in 1929 in [149] by von Neumann in relation with his investigations upon the Banach-Tarski paradox. Since then they have been the subject of an intensive study from many different perspectives with recent applications even in game theory [23, 26]. More information about this topic can be found in the monographs [77, 122].

Proposition 2.3.1 *Amenable groups are sofic.*

Proof Suppose that Γ is an amenable countable discrete group, F is a finite subset of Γ, and ε is a positive real number. Fix a finite $(F \cup F^{-1}, \varepsilon)$-invariant subset K of Γ. If $\gamma \in F$ then define

$$\sigma_\gamma(x) = \gamma x$$

for $x \in \gamma^{-1}K \cap K$ and extended σ_γ arbitrarily to a permutation of K. Observe that this defines an $(F, 2\varepsilon)$-approximate morphism from G to the group S_K of permutations of K endowed with the Hamming length (which is isomorphic as invariant length group to S_n where n is the cardinality of K). This concludes the proof that Γ is sofic.   □

If C is a class of (countable, discrete) groups, then a group Γ is *locally embeddable* into elements of C if for every finite subset F of Γ there is a function Φ from Γ to a group $T \in C$ such that Φ is nontrivial and preserves the operation on F, i.e. for every $g, h \in F$:

$$\Phi(gh) = \Phi(g)\Phi(h),$$

and

$$\Phi(g) \neq 1, \qquad \text{for all } g \neq 1.$$

The group Γ is *residually* in C if moreover Φ is required to be a surjective *-homomorphism. Recall also that Γ is *locally* in C if every finitely generated subgroup of Γ belongs to C. It is clear that if Γ is either locally in C or residually in C, then in particular Γ is locally embeddable into elements of C.

It is clear from the very definition that soficity and hyperlinearity are property concerning only *finite subsets* of the group. This is made precise in Exercise 2.3.2.

Exercise 2.3.2 Show that a group that is locally embeddable into sofic groups is sofic. The same is true replacing sofic with hyperlinear.

In particular Exercise 2.3.2 shows that locally sofic and residually sofic groups are sofic.

Groups that are locally embeddable into finite groups were introduced and studied in [75] under the name of LEF groups. Since finite groups are sofic, Exercise 2.3.2 implies that LEF groups are sofic. In particular residually finite groups are sofic. More generally groups that are locally embeddable into amenable groups (LEA)—also called *initially subamenable*—are sofic. It is a standard result in group theory that free groups are residually finite (see [126, Example 1.3]). Therefore the previous discussion implies that free groups are sofic . Examples of hyperlinear and sofic groups that are not LEA have been recently constructed by Andreas Thom [143] and Yves de Cornulier [45], respectively. We will present these examples in Sect. 2.5.

It should be now mentioned that it is to this day not know if there is any group which is not sofic, nor if there is any group which is not hyperlinear.

Open Problem 2.3.3 Are the classes of sofic and hyperlinear groups proper subclasses of the class of all countable discrete groups?

It is not even known if every *one-relator group*, i.e. a finitely presented group with only one relation, is necessarily sofic or hyperlinear. This is an open problem suggested by Nate Brown. Similarly it is not known whether all *Gromov hyperbolic* groups are sofic. In fact it is not known whether there is a Gromov hyperbolic groups that is not residually finite. An example of a *monoid* that does not satisfy the natural generalization of soficity for monoids is provided in [32].

2.4 Closure Properties of the Classes of Sofic and Hyperlinear Groups

The classes of sofic and hyperlinear groups have nice closure properties.

Proposition 2.4.1 *The class of sofic groups is closed with respect to the following operations:*

1. *Subgroups;*
2. *Direct limits;*
3. *Direct products;*
4. *Inverse limits;*
5. *Extensions by amenable groups;*
6. *Free products;*
7. *Free products amalgamated over an amenable group;*

8. HNN extension over an amenable group;
9. Graph product.

It is clear from the definition that soficity is a local property. Therefore a group is sofic if and only if all its (finitely generated) subgroups are sofic. In particular subgroups of sofic groups are sofic. It immediately follows that the direct limit of sofic groups is sofic.

To see that the direct product of two sofic groups is sofic, suppose that Γ_0, Γ_1 are sofic groups and F_0, F_1 are finite subsets of Γ_0 and Γ_1 respectively. If σ_0 and σ_1 are elements of S_n and S_m respectively, then define $\sigma_0 \otimes \sigma_1 \in S_{nm}$ by

$$(\sigma_0 \otimes \sigma_1)\,(im + j) = \sigma_0(i)m + \sigma_1\,(j)$$

for $i \in n$ and $j \in m$. If $\Phi_0 : \Gamma_0 \rightarrow S_n$ is an (F_0, ε)-approximate morphism and $\Phi_1 : \Gamma_1 \rightarrow S_m$ is an (F_1, ε)-approximate morphism, then the map

$$\Phi_0 \otimes \Phi_1 : \Gamma_0 \times \Gamma_1 \rightarrow S_{nm}$$

defined by

$$(\gamma_0, \gamma_1) \mapsto \Phi_0\,(\gamma_0) \otimes \Phi_1\,(\gamma_1)$$

is an $(F_0 \times F_1, 2\varepsilon)$-approximate morphism. This observation is sufficient to conclude that a direct product of two sofic groups is sofic. The result easily generalizes to arbitrary direct products in view of the local nature of soficity. Since the inverse limits is a subgroup of the direct product, it follows from what we have observed so far that the inverse limit of sofic groups is sofic.

We now prove the result—due to Elek and Szabó [55]—that the extension of a sofic group by an amenable group is sofic. Suppose that Γ is a group, and N is a normal sofic subgroup of Γ such that the quotient Γ/N is amenable. We want to show that Γ is sofic. Fix $\varepsilon > 0$ and a finite subset F of Γ. Denote by $g \mapsto \bar{g}$ the canonical quotient map from Γ to Γ/N and let r be a (set-theoretic) inverse for the quotient map. Observe that $r\,(\bar{g})^{-1}\,g \in N$ for every $g \in G$. Følner's reformulation of amenability yields a finite subset $\overline{A}$ of $\Gamma \backslash N$ such that

$$\left|\overline{A}\bar{g}\backslash\overline{A}\right| \leq \varepsilon\,\left|\overline{A}\right|$$

for every $g \in F$. Let A be the image of $\overline{A}$ under r and $\hat{F}$ be $N \cap (A \cdot F \cdot A^{-1})$. Since N is sofic there is a $(\hat{F}, \varepsilon)$-approximate morphism $g \mapsto \sigma_g$ from N to S_n for some $n \in \mathbb{N}$. Let k be $n|A| = n\,|\overline{A}|$ and identify S_k with the group of permutations of $n \times A$. Define the map $g \mapsto \tau_g$ from Γ to S_k by setting

$$\tau_g\,(i, h) = \begin{cases} \left(\sigma_{\overline{r(gh)}^{-1}gh}\,(i)\,, r(\overline{gh})\right) & \text{if } \overline{gh} \in \overline{A} \\ (i, h) & \text{otherwise.} \end{cases}$$

This is well defined since, as we observed above, $r\left(\overline{gh}\right)^{-1}gh \in N$.

Exercise 2.4.2 Show that the map $g \mapsto \tau_g$ is an $(F, 3\varepsilon)$-approximate morphism from Γ to S_k.

Hint Suppose that $g \in F \backslash \{1_\Gamma\}$. To show that τ_g has at most $\varepsilon n |A|$ fixed points, distinguish the cases when $g \notin N$ and $g \in N$. In the first case observe that (i, h) is a fixed point only if $\overline{gh} \notin \overline{A}$, and use the fact that $|\overline{A}\overline{g} \backslash \overline{A}| \leq \varepsilon |\overline{A}|$. In the second case observe that (i, h) is a fixed point only if i is a fixed point for $\sigma_{h^{-1}gh}$, and use the fact that $h^{-1}gh \in \hat{F}$ and $\gamma \mapsto \sigma_\gamma$ is a $(\hat{F}, \varepsilon)$-approximate morphism. To conclude fix $g, g' \in F$ and observe that for all but $\varepsilon |A|$ values of $h \in A$, $\overline{gh}$ and $\overline{g'gh}$ both belong to $\overline{A}$. For such values of h show that for all but εn values of $i \in n$ one has that $\tau_{g'}\tau_g (i, h) = \tau_{g'g} (i, h)$. ∎

It is a little more involved to observe that the free product of sofic groups is sofic, which is a result of Elek and Szabó from [55]. Suppose that Γ_0, Γ_1 are sofic groups, F_0, F_1 are finite subsets of Γ_0 and Γ_1, and $\varepsilon > 0$. We identify canonically Γ_0 and Γ_1 with subgroups of the free product $\Gamma = \Gamma_0 * \Gamma_1$. Every element γ of Γ has a unique shortest decomposition

$$\gamma = g_1 h_1 \cdots g_n h_n$$

with $g_i \in \Gamma_0$ and $h_i \in \Gamma_1$. Fix $N \in \mathbb{N}$ and set F to be the set of elements of γ of Γ such that the unique shortest decomposition of γ has length at most N and its elements belong to $F_0 \cup F_1$. By Exercise 2.1.10 one can find $n \in \mathbb{N}$, an (F_0, ε)-approximate morphism $g \mapsto \sigma_g^{(0)}$ from Γ_0 to S_n and an (F_1, ε)-approximate morphism $g \mapsto \sigma_g^{(1)}$ from Γ_1 to S_n such that $\left(\sigma_g^{(i)}\right)^{-1} = \sigma_{g^{-1}}^{(i)}$ and $\sigma_g^{(i)}$ has no fixed points for $g \in F^{(i)} \backslash \{1_{\Gamma_i}\}$ and $i = 0, 1$. It follows from the already recalled fact that free groups are residually finite that there is a finite group V generated by elements v_{ij} for $i, j \in [n]$ such that the generators satisfy no nontrivial relation expressed by words of length at most N. Equivalently the Cayley graph of V with respect to the generators v_{ij} has no cycle of length smaller than or equal to N. Consider then $k = n^2 |V|$ and identify S_k with the permutation group of $n \times n \times V$. Define for $g \in \Gamma_0$ the permutation τ_g of $n \times n \times V$ by

$$\tau_g (i, j, w) = \left(\sigma_g^{(0)} (i), j, w\right).$$

Similarly define for $h \in \Gamma_1$ the permutation τ_h of $[n] \times [n] \times V$ by

$$\tau_h (i, j, w) = \left(i, \sigma_h^{(1)} (j), w \cdot v_{ij}^{-1} \cdot v_{i, \sigma_h^{(1)}(j)}\right).$$

Finally is γ is an element of F with shortest decomposition $g_1 h_1 \cdots g_n h_n$ for $n \leq N$ set

$$\tau_\gamma = \tau_{g_1} \circ \tau_{h_1} \circ \cdots \circ \tau_{g_n} \circ \tau_{h_n}.$$

Exercise 2.4.3 Show that the map $\gamma \mapsto \tau_\gamma$ defied above is an (F, ε)-approximate morphism from Γ to S_k.

Hint The fact that the generators v_{ij} of V satisfy no nontrivial relations expressed by words of length at most N implies that τ_γ has no fixed points whenever γ is a nonidentity element of F. Suppose now that $\gamma, \gamma' \in F \setminus \{1\}$ and consider the corresponding unique shortest decompositions $\gamma = g_1 h_1 \cdots g_n h_n$ and $\gamma' = g'_1 h'_1 \cdots g'_n h'_n$. In order to show that $\tau_\gamma \circ \tau_{\gamma'}$ and $\tau_{\gamma\gamma'}$ differ on at most εk points of $[n] \times [n] \times V$ consider different cases depending whether both, none, or exactly one between h_n and g'_1 is equal to the identity. ∎

The result about the free product of sofic groups has been later generalized by Collins and Dykema, who showed that the free product of sofic groups amalgamated over a monotileable amenable group is sofic [42]. The monotileability assumption has been later removed by Elek and Szabó [56, Theorem 1] and, independently, by Paunescu [123].

Suppose that Γ is a group with presentation $\langle S|R \rangle$, H is a subgroup of Γ, and $\alpha : H \to \Gamma$ is an injective homomorphism. Let t be a symbol not in Γ. The corresponding *HNN extension* $\hat{\Gamma}$ is the group generated by $S \cup \{t\}$ subject to the relations R and $t^{-1} h t = \alpha(h)$ for $h \in H$. We claim that $\hat{\Gamma}$ is sofic whenever Γ is sofic an H is amenable. Define $\Gamma_i = t^{-i} \Gamma t^i$ for $i \in \mathbb{Z}$, and let S be the subgroup of $\hat{\Gamma}$ generated by Γ_i for $i \in \mathbb{Z}$. Then $\hat{\Gamma}$ is an extension of S by $\mathbb{Z}$. In particular to show that $\hat{\Gamma}$ is sofic it suffices to show that S is sofic. To show that S is sofic it is enough to show that for every $n \in \mathbb{N}$ the group $S_n = \langle \Gamma_i : i \in [-n, n] \cap \mathbb{Z} \rangle$ is sofic. This can be shown by induction on n observing that $S_0 = \Gamma$, and S_{n+1} can be obtained as a free product of S_n and isomorphic copies of Γ amalgamated over amenable subgroups isomorphic to H.

A different generalization of the result about free products consists in considering the graph products of sofic groups. Suppose that (V, E) is a simple undirected graph and, for every $v \in V$, Γ_v is a group. The corresponding graph product is the quotient of the free products of the Γ_v's by the normal subgroup generated by the relators $[g_v, g_w]$ for $g_v \in \Gamma_v$ and $g_w \in \Gamma_w$ such that v and w are connected by an edge. Clearly free products correspond to graph products where there are no edges. Theorem 1.1 of [38] refines the argument for free products to show that in fact an arbitrary graph product of sofic groups is sofic.

2.5 Further Examples of Sofic Group

In this section we want to list some interesting examples of sofic and hyperlinear groups. It follows from the discussion in Sect. 2.3 that all residually finite groups are sofic. In particular free groups are sofic. To obtain an example of a sofic group that is not residually finite one can consider, for $n, m \geq 2$ distinct, the Baumslag-Solitar group $BS(m, n)$ generated by two elements a, b satisfying the relation $a^{-1} b^m a = b^n$. These groups were introduced in [11] to provide examples of simple finitely presented groups that are not Hopfian, i.e. admit a surjective but not injective endomorphism. It is known that one hand such groups are not residually finite [113, Theorem C]. On the other hand they are residually solvable [104, Corollary 2] and, in particular, sofic.

An example of a sofic (and in fact LEF) finitely generated group that is not residually amenable is provided in [55, Theorem 3], adapting a construction from [75]. Recall that a function $\phi : \Gamma \to \mathbb{C}$ has positive type if whenever $\lambda_i \in \mathbb{C}$ and $g_i \in \Gamma$ are for $i \leq n$ then

$$\sum_{i,j \leq n} \lambda_i \overline{\lambda}_j \phi\left(g_i^{-1} g_j\right) \geq 0.$$

A group Γ has Kazhdan's property (T) if any sequence of positive type functions on G that converges pointwise to the function $\mathbf{1}$ constantly equal to 1 in fact converges to $\mathbf{1}$ uniformly. Such a property, originally introduced by Kazhdan in [90], has played a key role in the latest developments of geometric group theory; see [12]. Property (T) can be regarded as a strong antithesis to amenability, since the only amenable groups with property (T) are the finite groups. Since property (T) is preserved by quotients, it follows that an amenable quotient of a property (T) group must be finite. We also recall here that a group with property (T) is finitely generated [12, Theorem 1.3.1].

Consider an infinite hyperbolic residually finite property (T) group K. (Examples of such groups are provided in [78].) Let P be the group of finitely supported permutations of K, and Q be the group generated by P and the left translations by elements of K. Then P is a normal subgroup of Q and Q is a semidirect product of P and K. Consistently we identify K with a subgroup of Q. Let S be a finite set of generators for K and T_S be the set of transpositions of the form $(1, s)$ for $s \in S$.

Exercise 2.5.1 Show that $S \cup T_S$ is a set of generators for Q.

Hint Observe that if $s \in S$ and $g \in K$ then the transposition (g, gs) can be written as $g^{-1} (1, s) g$. Recall that the group of permutations on n symbols $x_1, \ldots, x_n$ is generated by the transpositions (x_i, x_{i+1}) for $i \leq n - 1$. ∎

We now observe that Q is not residually amenable. Recall that every hyperbolic group contains an nontorsion element. Let $t \in K$ such an element, and let $a \in P$ be the transposition $(t, 1)$. Observe that $t^n a t^{-n} = \left(t^{n+1}, t\right) \neq a$ for every $n \in \mathbb{N}$. Suppose by contradiction that there is a homomorphism $\phi : Q \to M$ into an amenable group M such that $\phi(a) \neq 1_M$ and $\phi(t) \neq 1_M$. Let A be the (simple) subgroup of P of even permutations. Since A is simple, $a \in A$, and $\phi(a) \neq 1_M$, ϕ must be injective on A. As observed before the fact that K has property (T) implies that the image of K under ϕ is finite. Let n be the rank of $\phi[K]$ and observe that $\phi(t^n) = 1$. Therefore $\phi(t^{-n} a t^{-n}) = \phi(a)$ while $t^{-n} a t^{-n} \neq a$. This contradicts the injectivity of ϕ on A.

We now show that Q is sofic and, in fact LEF. Denote by $B_n(K)$ the n-ball around 1_K in the Cayley graph of K associated with the generating set S. Define F_n to be the set of elements of Q of the form $k\sigma$ where $k \in B_n(K)$ and $\sigma \in P$ is supported on

$B_n(K)$. We want to define an injective map from F_{2n} to a finite group that preserves the operation on F_n. Since K is residually finite it has a finite-index normal subgroup N_n such that $N_n \cap B_{2n}(K) = \{1_K\}$. Define H_n to be the (finite) group of permutations on K/N_n. Denote by π_n the quotient map from K onto K/H_n and define $\psi_n : F_{2n} \to H_n$ by setting

$$\psi_n(k\sigma) = \psi_n(k)\,\psi_n(\sigma).$$

Here $\psi_n(k)$ is the left translation by $\pi(k)$, while $\psi_n(\sigma)$ is defined by $\psi_n(\sigma)(\pi_n(h)) = \pi(\sigma(h))$ if $h \in B_{2n}(K)$ and acts as the identity otherwise. Clearly ψ_n is injective on F_{2n}.

Exercise 2.5.2 Show that $\psi_n(xy) = \psi_n(x)\,\psi_n(y)$ for $x, y \in F_n$.

Hint Want to show that $\psi_n(xy)(\pi_n(h)) = \psi_n(x)\,\psi_n(y)(\pi_n(h))$ for every $h \in K$. Distinguish the cases when $h \in B_{2n}(K)$ and $h \notin B_{2n}(K)$. ∎

Another example of a not residually finite sofic group which is moreover finitely presented (and in fact one-relator) was provided by Jon Bannon in [8]; see also [9] for other similar examples. Consider the group Γ generated by two elements a, b subject to the relation $a = [a, a^b]$ where $a^b = bab^{-1}$ and $[a, a^b]$ is the commutator of a and a^b. Such a group was introduced by Baumslag in [10] as an example of a non-cyclic one-relator group all of whose finite quotients are cyclic. In particular Γ is not residually finite and, in fact, not residually solvable since a belongs to all the derived subgroups of Γ. We now show that it is sofic.

Set $x = a^{-1}$ and $y = bab^{-1}$. Then the relator $a^{-1}[a, a^b]$ becomes $x^{-2}y^{-1}xy$. Observe that the group $H = \langle x, y | y^{-1}xy = x^2 \rangle$ is the Baumslag-Solitar group $B(1, 2)$ and, in particular, sofic. Moreover Γ is the HNN extension of H with respect to the isomorphism $x^n \mapsto y^n$ between the subgroups $\langle x \rangle$ and $\langle y \rangle$ of H. It follows that Γ is sofic since HNN extensions of sofic groups over amenable groups are sofic.

We would now like to present examples of hyperlinear and sofic groups that are not LEA. Recall that a group is locally embeddable into amenable groups (LEA) if, briefly, every finite portion of its multiplication table can be realized as a portion of the multiplication table of an amenable group. The first example of a hyperlinear not LEA group was constructed by Thom in [143], adapting a construction due to de Cornulier [47] and Abels [1].

If $n, m \in \mathbb{N}$ denote by $M_{n,m}\left(\mathbb{Z}\left[\frac{1}{p}\right]\right)$ the space of $n \times m$ matrices over $\mathbb{Z}\left[\frac{1}{p}\right]$ and by $SL_n\left(\mathbb{Z}\left[\frac{1}{p}\right]\right)$ the group of $n \times n$ matrices of determinant 1. Let Γ be the group of matrices in $SL_8\left(\mathbb{Z}\left[\frac{1}{p}\right]\right)$ of the form

$$\begin{bmatrix} 1 & a_{12} & a_{13} & a_{14} \\ 0 & a_{22} & a_{23} & a_{24} \\ 0 & 0 & a_{33} & a_{34} \\ 0 & 0 & 0 & 1 \end{bmatrix}$$

where the diagonal blocks are square matrices of rank $1, 3, 3, 1$ and $a_{22}, a_{33} \in SL_3\left(\mathbb{Z}\left[\frac{1}{p}\right]\right)$. The center C of Γ consists of the matrices of the form

$$\begin{bmatrix} 1 & 0 & 0 & a \\ 0 & I & 0 & 0 \\ 0 & 0 & I & 0 \\ 0 & 0 & 0 & 1 \end{bmatrix}$$

where I is the 3×3 identity matrix and $a \in K$. Thus C is isomorphic to the additive group of $\mathbb{Z}\left[\frac{1}{p}\right]$. In particular C contains a subgroup Z isomorphic to $\mathbb{Z}$.

It is shown in [47, Proposition 2.7] that Γ is a lattice in a locally compact group with property (T). Since property (T) passes to lattices [12, Theorem 1.7.1], it follows that Γ has property (T) and, in particular, is finitely generated. It is moreover shown in [47, Sect. 3] that Γ and (hence) Γ/Z are finitely presented. We present here the argument to show that Γ/Z is not Hopfian, i.e. it has a surjective endomorphism with nontrivial kernel.

Denote by $\mathbb{Z}\left[\frac{1}{p}\right]^{\times}$ the multiplicative group of invertible elements of $\mathbb{Z}\left[\frac{1}{p}\right]$. Then $\mathbb{Z}\left[\frac{1}{p}\right]^{\times}$ identified with the group of matrices of the form

$$\begin{bmatrix} a & 0 & 0 & 0 \\ 0 & 1 & 0 & 0 \\ 0 & 0 & 1 & 0 \\ 0 & 0 & 0 & 1 \end{bmatrix}$$

naturally acts on Γ by conjugation. Considering $p \in \mathbb{Z}\left[\frac{1}{p}\right]^{\times}$ gives an automorphism β of Γ that maps the center Z of Γ to its property subgroup Z^p. Thus β induces a surjective endomorphism $\overline{\beta}$ of Γ/Z whose kernel is the (nontrivial) subgroup C/Z of Γ/Z; see [47, Lemma 2.3]. As a consequence Γ/Z is not Hopfian.

As observed in [143, 3.1] a finitely presented LEA group that has property (T) is necessarily Hopfian. In fact a finitely presented LEA group is residually amenable by Ceccherini-Silberstein and Coornaert [31, Proposition 7.3.8]. As recalled before an amenable quotient of a property (T) group is necessarily finite. Therefore a property (T) residually amenable group is residually finite. Finally a residually finite finitely generated group is easily seen to be Hopfian. It follows from this argument that the group Γ/Z, being finitely presented, property (T), and not Hopfian, is *not* LEA.

To conclude one needs to observe that Γ/Z is hyperlinear. Considering the reduction modulo q where q is an arbitrary positive integer prime with p shows that the group Γ is residually finite and, in particular, hyperlinear. The argument is concluded by showing that hyperlinearity is preserved by taking quotients by central subgroups; see [143, Remark 3.4].

It is not known whether the group Γ/Z above is sofic. We conclude this sequence of examples with an example due to de Cornulier of a finitely presented sofic group that is not LEA [45]. Let Γ be the group of matrices

$$\begin{bmatrix} a & b & u_{02} & u_{03} & u_{04} \\ c & d & u_{12} & u_{13} & u_{14} \\ 0 & 0 & p^{n_2} & u_{23} & u_{24} \\ 0 & 0 & 0 & p^{n_3} & u_{34} \\ 0 & 0 & 0 & 0 & 1 \end{bmatrix}$$

with

- $\begin{bmatrix} a & b \\ c & d \end{bmatrix} \in SL_2(\mathbb{Z})$,
- $u_{ij} \in \mathbb{Z}\left[\frac{1}{p}\right]$, and
- $n_2, n_3 \in \mathbb{Z}$.

Let M be the normal subgroup of Γ consisting of matrices of the form

$$\begin{bmatrix} 1 & 0 & 0 & 0 & m_1 \\ 0 & 1 & 0 & 0 & m_2 \\ 0 & 0 & 1 & 0 & 0 \\ 0 & 0 & 0 & 1 & 0 \\ 0 & 0 & 0 & 0 & 1 \end{bmatrix}$$

with $m_1, m_2 \in \mathbb{Z}\left[\frac{1}{p}\right]$. The normal subgroup $M_{\mathbb{Z}}$ is defined similarly with $m_1, m_2 \in \mathbb{Z}$. Then one can consider the quotient $\Gamma/M_{\mathbb{Z}}$. It is shown in [45] that such a group is sofic and finitely presented but not LEA.

In order to prove soficity, it is enough to show that Γ can be obtained as an extension of a sofic group by an amenable group. Consider the normal subgroup Υ of elements of Γ for which $n_2 = n_3 = 0$. Observe that Γ/Υ is isomorphic to $\mathbb{Z}^2$. Therefore it remains to show that $\Upsilon/M_{\mathbb{Z}}$ is sofic.

Fix $m \in \mathbb{N}$ and consider the subgroup Υ_m of elements of Υ for which

- $u_{02}, u_{12}, u_{23}, u_{34} \in p^{-m}\mathbb{Z}$,
- $u_{03}, u_{13}, u_{24} \in p^{-2m}\mathbb{Z}$, and
- $u_{04}, u_{14} \in p^{-3m}\mathbb{Z}$.

It is clear that $\bigcup_m \Upsilon_m = \Upsilon$. Therefore by the local nature of soficity we are left with the problem of showing that $\Upsilon_m/M_{\mathbb{Z}}$ is sofic for every $m \in \mathbb{N}$. Such a group is in fact residually finite and hence sofic. Consider the subgroup Λ of elements of Γ for which the block

$$\begin{bmatrix} a & 0 \\ 0 & b \end{bmatrix} \in SL_2(\mathbb{Z})$$

is the identity matrix. Observe that $(\Upsilon_m \cap \Lambda)/M_\mathbb{Z}$ is a normal subgroup of $\Upsilon_m/M_\mathbb{Z}$, and moreover $\Upsilon_m/M_\mathbb{Z}$ is isomorphic to the semidirect product $(\Upsilon_m \cap \Lambda)/M_\mathbb{Z} \rtimes SL_2(\mathbb{Z})$. Now $\Upsilon_m \cap \Lambda$ is solvable and finitely generated, and hence residually finite [80, 2.1]. The proof is concluded by observing that a semidirect product of a finitely generated residually finite group by a residually finite group is residually finite.

Observe that the quotient map $\Gamma \mapsto \Gamma/M_\mathbb{Z}$ restricted to the subgroup (isomorphic to $SL_2(\mathbb{Z})$) of elements

$$\begin{bmatrix} a & b & 0 & 0 & 0 \\ c & d & 0 & 0 & 0 \\ 0 & 0 & 1 & 0 & 0 \\ 0 & 0 & 0 & 1 & 0 \\ 0 & 0 & 0 & 0 & 1 \end{bmatrix}$$

is injective. Thus $\Gamma/M_\mathbb{Z}$ contains a copy of $SL_2(\mathbb{Z})$ and, hence, of the free group on 2 generators F_2. (Recall that the matrices

$$A = \begin{bmatrix} 1 & 2 \\ 0 & 1 \end{bmatrix} \quad \text{and} \quad B = \begin{bmatrix} 1 & 0 \\ 2 & 1 \end{bmatrix}$$

generate a copy of F_2 inside $SL_2(\mathbb{Z})$; see [126, page 3].)

By [31, Corollary 7.1.20] in order to show that $\Gamma/M_\mathbb{Z}$ is not LEA it is enough to show that it is an isolated point in the space of marked groups. Fix $n \in \mathbb{N}$. An n-marked group is a group Γ endowed with a distinguished generating n-tuple $(\gamma_1, \ldots, \gamma_n)$. Let $\mathcal{G}_n$ be the space of n-marked groups. Given an n-marked group one can consider the kernel of the epimorphism $F_n \to \Gamma$ mapping the canonical free generators of F_n to the γ_i's. Conversely any normal subgroup N of F_n gives rise to the n-marked group F_n/N where the distinguished n-tuple of generators is the image of the free generators of F_n. This argument shows that one can identify the space $\mathcal{G}_n$ of n-marked group with the space of the normal subgroups of F_n. Identifying in turn a normal subgroup of F_n with its characteristic function yields an inclusion of $\mathcal{G}_n$ into 2^{F_n} as a closed subspace. This defines a compact metrizable zero-dimensional topology on $\mathcal{G}_n$. Corollary 7.1.20 in [31] shows that a group is LEA if and only if it is a limit of amenable groups in the space of marked groups. Therefore since $\Gamma/M_\mathbb{Z}$ is not LEA, it is enough to show that $\Gamma/M_\mathbb{Z}$ is *isolated*, i.e. it is an isolated point in the space of marked groups.

It is shown in [46, Lemma 1] that being isolated is indeed a well defined property of a group, independent of the marking. Proposition 2 in [46] provides the following characterization of being isolated: a group is isolated if and only if it is finitely presented and moreover finitely discriminable, i.e. it contains a finite subsets that meets every nontrivial normal subgroup in a nonidentity element. The proof that $\Gamma/M_\mathbb{Z}$ satisfies these conditions is presented in [45, Sect. 3] and [46, 5.4].

Other examples of finitely presented sofic groups that are not LEA are provided in [94].

2.6   Logic for Invariant Length Groups

The logic for metric structures is a generalization of the usual first order logic. It is a natural framework to study algebraic structures endowed with a nontrivial metric and their elementary properties (i.e. properties preserved by ultrapowers or equivalently expressible by formulae). In the sequel we introduce a particular instance of logic for metric structures to describe and study groups endowed with an invariant length functions.

A *term* $t(x_1, \ldots, x_n)$ in the language of invariant length groups in the variables $x_1, \ldots, x_n$ is a *word* in the indeterminates $x_1, \ldots, x_n$, i.e. an expression of the form

$$x_{i_1}^{n_1} \ldots x_{i_l}^{n_l}$$

for $l \in \mathbb{N}$ and $n_i \in \mathbb{Z}$ for $i = 1, 2, \ldots, l$. For example

$$xyx^{-1}y^{-1}$$

is a term in the variables x, y. The empty word will be denoted by 1. If G is an invariant length group, $g_1, \ldots, g_m$ are elements of G, and $t(x_1, \ldots, x_n, y_1, \ldots, y_m)$ is a term in the variables $x_1, \ldots, x_n, y_1, \ldots, y_m$, then one can consider the term $t(x_1, \ldots, x_n, g_1, \ldots, g_m)$ with *parameters* from G, which is obtained from $t(x_1, \ldots, x_n, y_1, \ldots, y_m)$ replacing formally y_i with g_i for $i = 1, 2, \ldots, m$. The evaluation $t^G(x_1, \ldots, x_n)$ in a given invariant length group G of a term $t(x_1, \ldots, x_n)$ in the variables $x_1, \ldots, x_n$ (possibly with parameters from G) is the function from G^n to G defined by

$$(g_1, \ldots, g_n) \mapsto t(g_1, \ldots, g_n)$$

where $t(g_1, \ldots, g_n)$ is the element of G obtained replacing in $t(x_1, \ldots, x_n)$ every occurrence of x_i with g_i for $i = 1, 2, \ldots, n$. For example the evaluation in a invariant length group G of the term $xyx^{-1}y^{-1}$ is the function from G^2 to G that assigns to every pair (g, h) of elements of G their commutator $ghg^{-1}h^{-1}$. The evaluation of the empty word is the function on G constantly equal to 1_G.

A *basic formula* φ in the variables $x_1, \ldots, x_n$ is an expression of the form

$$\ell\left(t(x_1, \ldots, x_n)\right)$$

where $t(x_1, \ldots, x_n)$ is a term in the variables $x_1, \ldots, x_n$. The evaluation $\varphi^G(x_1, \ldots, x_n)$ of $\varphi(x_1, \ldots, x_n)$ in an invariant length group G is the function from G^n to $[0, 1]$ defined by

$$(g_1, \ldots, g_n) \mapsto \ell_G\left(t^G(g_1, \ldots, g_n)\right)$$

where ℓ_G is the invariant length of G. For example

$$\ell(xyx^{-1}y^{-1})$$

is a basic formula whose interpretation in an invariant length group G is the function assigning to a pair of elements of G the length of their commutator. This basic formula can be thought as measuring how much x and y commute. The evaluation at (g, h) of its interpretation in an invariant length group G will be 0 if and only if g and h commute.

Finally a *formula* φ is any expression that can be obtained starting from basic formulae, composing with continuous functions from $[0, 1]^n$ to $[0, 1]$, and taking infima and suprema over some variables. In this framework continuous functions have the role of *logical connectives*, while infima and suprema should be regarded as *quantifiers*. With these conventions the terminology from the usual first order logic is used in this setting. A formula is *quantifier-free* if it does not contain any quantifier. A variable x in a formula φ is *bound* if it is within the scope of a quantifier over x, and *free* otherwise. a formula without free variables is called a *sentence*. The interpretation of a formula in a length group G is defined in the obvious way by recursion on its complexity. For example

$$\sup_{x} \sup_{y} \ell(xyx^{-1}y^{-1})$$

is a sentence, with bound variables x and y. Its evaluation in an invariant length group G is the real number

$$\sup_{x \in G} \sup_{y \in G} \ell_G(xyx^{-1}y^{-1}).$$

This sentence can be thought as measuring how much the group G is abelian. Its interpretation in G is zero if and only if G is abelian. This example enlightens the fact that the possible truth values of a sentence (i.e. values of its evaluations in an invariant length group) are all real numbers between 0 and 1. Moreover 0 can be thought as "true" while strictly positive real numbers of different degrees as "false". In this spirit we say that a sentence φ holds in G if and only if its interpretation in G is zero. Using this terminology we can assert for example that an invariant length group G is abelian if and only if the formula

$$\sup_{x} \sup_{y} \ell(xyx^{-1}y^{-1})$$

holds in G. Observe that if φ is a sentence, then $1 - \varphi$ is a sentence such that φ holds in G if and only if the interpretation of $1 - \varphi$ in G is 1. Thus $1 - \varphi$ can be though as a sort of *negation* of the sentence φ. Another example of sentence is

$$\sup_{x} \min \{|\ell(x) - 1|, |\ell(x)|\}.$$

Such sentence holds in an invariant length group G if and only if the invariant length function in G attains values in $\{0, 1\}$, i.e. it is the trivial invariant length function on G. It is worth noting at this point that for any sentence φ as defined in the logic for invariant length groups there is a corresponding formula φ_0 in the usual (discrete) first order logic in the language of groups such that the evaluation of φ_0 in a discrete group G coincides with the evaluation of φ in G regarded as an invariant metric group endowed with the trivial invariant length function. For example the (metric) formula expressing that a group is abelian corresponds to the (discrete) formula

$$\forall x \forall y \, (xy = yx).$$

Sentences in the language of invariant length groups allow one to determine which properties of an invariant length group are elementary. A property concerning length groups is *elementary* if there is a set Φ of sentences such that an invariant length group has the given property if and only if G satisfies all the sentences in Φ. For example the property of being abelian is elementary, since an invariant length group is abelian if and only if it satisfies the sentence $\sup_{x,y} \ell(xyx^{-1}y^{-1})$. Two invariant length groups G and G' are *elementarily equivalent* if they have the same elementary properties. This amounts to say that any sentence has the same evaluation in G and G'. A class $\mathcal{C}$ of invariant length groups will is *axiomatizable* if the property of belonging to $\mathcal{C}$ is elementary. The previous example of sentence shows that the class of abelian length groups is axiomatizable by a single sentence. Elementary properties and classes are closely related with the notion of ultraproduct of invariant length groups.

Suppose that $(G_n)_{n \in \mathbb{N}}$ is a sequence of invariant length groups and $\mathcal{U}$ is a free ultrafilter over $\mathbb{N}$. The *ultraproduct* $\prod_{\mathcal{U}} G_n$ of the sequence $(G_n)_{n \in \mathbb{N}}$ with respect to the ultrafilter $\mathcal{U}$ is by definition quotient of the product $\prod_n G_n$ by the normal subgroup

$$N_{\mathcal{U}} = \left\{ (g_n)_{n \in \mathbb{N}} : \lim_{n \to \mathcal{U}} \ell_{G_n}(g_n) = 0 \right\}$$

endowed with the invariant length function

$$\ell_{\mathcal{U}}((g_n) N_{\mathcal{U}}) = \lim_{n \to \mathcal{U}} \ell_{G_n}(g_n).$$

As before a sequence $(g_n)_{n \in \mathbb{N}}$ in $\prod_n G_n$ is called *representative sequence* for the corresponding element in $\prod_{\mathcal{U}} G_n$. Observe that ultraproducts of the sequence $(S_n)_{n \in \mathbb{N}}$ of permutation groups endowed with the Hamming invariant length function as defined in Sect. 2.1 and ultraproducts of the sequence $(U_n)_{n \in \mathbb{N}}$ of unitary groups endowed with the Hilbert-Schmidt invariant length function as defined in Sect. 2.2 are particular cases of this definition. When the sequence $(G_n)_{n \in \mathbb{N}}$ is constantly equal to a fixed invariant length group G the ultraproduct $\prod_{\mathcal{U}} G_n$ is called *ultrapower* of G and denoted by $G^{\mathcal{U}}$. Observe that *diagonal embedding* of G into $G^{\mathcal{U}}$ assigning to $g \in G$ the element of $G^{\mathcal{U}}$ having the sequence constantly equal to g as representative sequence is a length preserving group homomorphism. This allows one to identify G with a subgroup of $G^{\mathcal{U}}$.

The ultraproduct construction behaves well with respect to interpretation of formulae. This is the content of a theorem proven in the setting of the usual first order logic by Łoś in [108]. Its generalization to the logic for metric structures can be found in [13] (Theorem 5.4). We state here the particular instance of Łoś' theorem in the context of invariant length groups.

Theorem 2.6.1 (Łoś) *Suppose that $\varphi(x_1, \ldots, x_k)$ is a formula with free variables $x_1, \ldots, x_k$, $(G_n)_{n \in \mathbb{N}}$ is a sequence of invariant length groups, and $\mathcal{U}$ is a free ultrafilter over $\mathbb{N}$. If $g^{(1)}, \ldots, g^{(k)}$ are elements of $\prod_n G_n$ then*

$$\varphi^{\prod_{\mathcal{U}} G_n}\left(g^{(1)}, \ldots, g^{(k)}\right) = \lim_{n \to \mathcal{U}} \varphi^{G_n}\left(g_n^{(1)}, \ldots, g_n^{(k)}\right)$$

where $g^{(i)}$ is any representative of the sequence $\left(g_n^{(i)}\right)$, for $i = 1, 2, \ldots, k$. In particular if φ is a sentence then

$$\varphi^{\prod_{\mathcal{U}} G_n} = \lim_{n \to \mathcal{U}} \varphi^{G_n}.$$

Theorem 2.6.1 can be proved by induction on the complexity of the formula φ. The particular instance of Theorem 2.6.1 when the sequence $(G_n)_{n \in \mathbb{N}}$ is constantly equal to an invariant metric group G shows that G and any ultrapower $G^{\mathcal{U}}$ of G are elementarily equivalent. Moreover the diagonal embedding of G into $G^{\mathcal{U}}$ is an *elementary embedding*, i.e. it preserves the value of formulae possibly with parameters from G.

A particularly useful property of ultraproducts is usually referred to as *countable saturation*. Roughly speaking countable saturation of a group H asserts that whenever one can find elements of H approximately satisfying any finite subset of a given countably infinite set of conditions, then in fact one can find an element of H exactly satisfying simultaneously all the conditions. In order to precisely define this property, and prove it for ultraproducts, we need to introduce some model-theoretic terminology, in the particular case of invariant length groups.

Definition 2.6.2 Suppose that H is a group endowed with an invariant length function. A countable set of formulae $\mathcal{X}$ in the free variables $x_1, \ldots, x_n$ and possibly with parameters from H is:

- *approximately finitely satisfiable* in H or *consistent* with H if for every $\varepsilon > 0$ and every finite collection of formulae $\varphi_1(x_1, \ldots, x_n), \ldots, \varphi_m(x_1, \ldots, x_n)$ from $\mathcal{X}$ there are $a_1, \ldots, a_n \in H$ such that

$$\varphi_i^H(a_1, \ldots, a_n) < \varepsilon$$

for every $i = 1, 2, \ldots, n$;
- *realized* in H if there are $a_1, \ldots, a_n \in H$ such that

$$\varphi^H(a_1, \ldots, a_n) = 0$$

for every formula $\varphi(x_1, \ldots, x_n)$ in $\mathcal{X}$. In this case the n-tuple (a) is called a realization of $\mathcal{X}$ in H.

In model-theoretic jargon, a set of formulae $\mathcal{X}$ as above is called a *type* over A, where A is the (countable) set of parameters of the formulae in $\mathcal{X}$. Clearly any set of formulae which is realized in H is also approximately finitely satisfiable H. The converse to this assertion is in general far from being true. For example suppose that H is the direct sum $G^{\oplus \mathbb{N}}$ of countably many copy of a countable group G with trivial center endowed with the trivial length function. Consider the set $\mathcal{X}$ of formulae

$$\max \left\{ |1 - \ell(x)|, \ell \left(xax^{-1}a^{-1} \right) \right\}$$

in the free variable x where a ranges over the elements of $G^{\oplus \mathbb{N}}$. Being the center of $G^{\oplus \mathbb{N}}$ trivial, the set of formulae $\mathcal{X}$ is not realized in $G^{\oplus \mathbb{N}}$. Nonetheless the fact that one can find in H a nonidentity element commuting with any given *finite* subset of $G^{\oplus \mathbb{N}}$ shows that $\mathcal{X}$ is approximately finitely satisfiable in H.

Definition 2.6.3 An invariant length group is *countably saturated* if any countable set of formulae $\mathcal{X}$ with parameters from H which is approximately finitely satisfied in H is realized in H.

Thus in a countably saturated invariant length group a countable set of formulae is approximately finitely satisfiable if and only if it is realized. Moreover it can be easily shown by recursion on the complexity that in the evaluation of a formula in a countably saturated structure infima and suprema can be replaced by minima and maxima respectively. More generally one can define κ-saturation for an arbitrary cardinal κ replacing countable types with types with less than κ parameters (it is not difficult to check that countable saturation is the same as $\aleph_1$-saturation). An invariant length group G is *saturated* if it is κ-saturated where κ is the *density character* of G, i.e. the minimum cardinality of a dense subset of G. The notions of saturation and countable saturation here introduced are the particular instances in the case of invariant length groups of the general model-theoretic notions of saturation and countable saturation. Being countably saturated is one of the fundamental features of ultraproducts. The proof of this fact in this context can be easily deduced from Łoś' theorem and is therefore left as an exercise.

Exercise 2.6.4 Suppose that $(G_n)_{n \in \mathbb{N}}$ is a sequence of invariant length groups, and $\mathcal{U}$ is a free ultrafilter over $\mathbb{N}$. Show that the ultraproduct $\prod_{\mathcal{U}} G_n$ is countably saturated.

Saturated structures have been intensively studied in model theory and have some remarkable properties. In the case of usual first order logic it is a consequence of the so called Chang-Makkai's theorem (see [34, Theorem 5.3.6]) that the automorphism group of a saturated structure of cardinality κ has 2^κ elements. The generalization to this result to the framework of logic for metric structures is an unpublished result of Ilijas Farah, Bradd Hart, and David Sherman. Proposition is the particular instance of such result in the case of invariant length groups of density character $\aleph_1$.

Proposition 2.6.5 *Suppose that G is an invariant length group. If G is countably saturated and has a dense subset of cardinality $\aleph_1$ then the group of automorphisms of G that preserve the length has cardinality $2^{\aleph_1}$.*

Proposition 2.6.6 is an immediate consequence of Proposition 2.6.5 and Exercise 2.6.4. Recall that the Continuum Hypothesis asserts that the cardinality of the power set of $\mathbb{N}$ coincides with the first uncountable cardinal $\aleph_1$. A cornerstone result in set theory asserts that the Continuum Hypothesis is independent from the usual axioms of set theory, i.e. can not be neither proved nor disproved (see [39, 40, 72]).

Proposition 2.6.6 *Suppose that* $(G_n)_{n\in\mathbb{N}}$ *is a sequence of separable invariant length groups. If the Continuum Hypothesis holds then* $\prod_{\mathcal{U}} G_n$ *has* $2^{\aleph_1}$ *length preserving automorphisms for any free ultrafilter* $\mathcal{U}$ *over* $\mathbb{N}$. *In particular* $\prod_{\mathcal{U}} G_n$ *has outer length-preserving automorphisms.*

Proof Assuming the Continuum Hypothesis $\prod_{\mathcal{U}} G_n$ has a dense subset of cardinality $\aleph_1$. Moreover by Exercise 2.6.4 $\prod_{\mathcal{U}} G_n$ is countably saturated. It follows from Proposition 2.6.5 that $\prod_{\mathcal{U}} G_n$ has $2^{\aleph_1}$ length-preserving automorphisms. $\square$

Corollary 2.6.7 is the particular instance of Proposition 2.6.6 when the sequence $(G_n)_{n\in\mathbb{N}}$ is the sequence of permutation groups or the sequence of unitary groups, one can obtain

Corollary 2.6.7 *Assume that the Continuum Hypothesis holds. For any free ultrafilter* $\mathcal{U}$ *over* $\mathbb{N}$ *the universal sofic group* $\prod_{\mathcal{U}} S_n$ *has* $2^{\aleph_1}$ *automorphisms. As a consequence* $\prod_{\mathcal{U}} S_n$ *has outer automorphisms. The same is true for the universal hyperlinear group* $\prod_{\mathcal{U}} U_n$.

We do not know if there exist models of set theory where some universal sofic or hyperlinear groups have only inner automorphisms. It is conceivable that this could be proved using ideas and methods from [136]. For example in [109] Lucke and Thomas show using a result from [136] that there is a model of set theory where some ultraproduct of the permutation groups *regarded as discrete groups* has only inner automorphisms.

Corollary 2.6.7 and the discussion that follows address a question of Păunescu from [124]: Theorem 4.1 and Theorem 4.2 of [124] show that all automorphisms of the universal sofic groups preserve the length function and the conjugacy classes. Păunescu then asks if it is possible that all automorphisms of the universal sofic groups are in fact inner. Corollary 2.6.7 in particular implies that such assertion does not follow from the usual axioms of set theory.

An application of Theorem 2 from [50] (known as Dye's theorem on automorphisms of unitary groups of factors) allows one to prove the analogue of Theorem 4.1 and Theorem 4.2 from [124] in the case of universal hyperlinear groups. More precisely all automorphisms of the universal hyperlinear groups preserve the length function, while the normal subgroup of automorphisms that preserves the conjugacy classes has index 2 inside the group of all automorphisms. The not difficult details can be found in [110].

Finally let us point out another consequence of countable saturation of ultraproducts. Suppose that Γ is a discrete group of size $\aleph_1$ such that every countable (or equivalently finitely generated) subgroup of Γ is sofic. It is not difficult to infer from Exercise 2.1.9 and countable saturation of $\prod_{\mathcal{U}} S_n$ that for any ultrafilter $\mathcal{U}$ over

$\mathbb{N}$ there is a length-preserving homomorphism from Γ (endowed with the trivial length function) to $\prod_{\mathcal{U}} S_n$. For example this implies that $\prod_{\mathcal{U}} S_n$ contains a free group on uncountably many generators. Analogue facts hold for hyperlinear groups and ultraproducts $\prod_{\mathcal{U}} U_n$.

2.7 Model Theoretic Characterization of Sofic and Hyperlinear Groups

In this section we shall show that the classes of sofic and hyperlinear groups are axiomatizable in the logic for invariant metric groups. Equivalently the properties of being sofic or hyperlinear are elementary. Recall that the quantifiers in the logic for invariant metric groups are inf and sup. More precisely sup can be regarded as the *universal quantifier*, analogue to $\forall$ in usual first order logic, while inf can be seen as the *existential quantifier*, which is denoted by $\exists$ in the usual first order logic. A formula is therefore called *universal* if it only contains universal quantifiers, and no existential quantifiers. The notion of *existential sentence* is defined in the same way. A *quantifier-free formula* is just a formula without any quantifier. Say that two formulae φ and φ' are *equivalent* if they have the same interpretation in any invariant length group. It can be easily proved by induction on the complexity that any universal sentence is equivalent to a formula of the form

$$\sup_{x_1} \ldots \sup_{x_n} \psi(x_1, \ldots, x_n)$$

where $\psi(x_1, \ldots, x_n)$ is quantifier-free. An analogous fact holds for existential sentences. It is easy to infer from this that if φ is a universal sentence, then $1 - \varphi$ is equivalent to an existential sentence, and vice versa. Exercise 2.7.1 together with Łoś' theorem on ultraproducts shows that universal and existential formulae have the same values in any ultraproduct of the symmetric groups regarded as invariant length groups.

Exercise 2.7.1 If φ is a universal sentence then the sequence $\left(\varphi^{S_n}\right)_{n \in \mathbb{N}}$ of its evaluation in the symmetric groups converges. Infer that the same is true for existential formulae. Deduce that the same holds for existential formulae.

Proof Fix $n \in \mathbb{N}$ and $\varepsilon > 0$. If $N > n$ write

$$N = kn + r$$

for $k \in \mathbb{N}$ and $r \in n$. Define the map $\iota : S_n \to S_N$ by

$$\iota(\sigma)(ik + j) = \begin{cases} in + \sigma(j) & \text{if } i \in k \text{ and } j \in n \\ in + j & \text{otherwise.} \end{cases}$$

Observe that ι is a group homomorphism that preserves the Hamming length function up to $\frac{1}{k}$. This means that

$$\left|\ell_{S_n}(\sigma) - \ell_{S_N}(\iota(\sigma))\right| \leq \frac{1}{k}.$$

Deduce that if φ is a universal sentence and $\varepsilon > 0$ then

$$\left|\varphi^{S_N} - \varphi^{S_n}\right| < \varepsilon.$$

for $N \in \mathbb{N}$ large enough.   □

Theorem 5.6 of [111] asserts that the same conclusion of Exercise 2.7.1 holds for formulae with alternation of at most two quantifiers. It is currently an open problem if the same holds for all formulae. A positive answer to this question would imply that the universal sofic groups, i.e. the length ultraproducts of the permutation groups, are pairwise isomorphic as invariant length groups if the Continuum Hypothesis holds. Theorem 1.1 of [145] asserts that if instead the Continuum Hypothesis fails there are $2^{2^{\aleph_0}}$ ultraproducts of the permutation groups that are pairwise nonisomorphic as discrete groups.

We can now show that the property of being sofic is axiomatizable. Suppose that Γ is a sofic group. By Exercise 2.1.9 there is a length preserving embedding of Γ into $\prod_{\mathcal{U}} S_n$ for any free ultrafilter $\mathcal{U}$, where Γ is endowed with the trivial length function. It is easily inferred from this that

$$\varphi^{\Gamma} \leq \varphi^{\prod_{\mathcal{U}} S_n} = \lim_{n \to +\infty} \varphi^{S_n}$$

for any universal sentence φ in the language of invariant length groups. If φ is instead an existential sentence then

$$\varphi^{\Gamma} \geq \lim_{n \to +\infty} \varphi^{S_n}.$$

In particular if φ is an existential sentence that holds in Γ, then $\lim_{n \to +\infty} \varphi^{S_n}$. Exercise 2.7.2 shows that this condition is sufficient for a group to be sofic.

Exercise 2.7.2 Suppose that Γ is a (countable discrete) group with the property that for any existential sentence φ that holds in Γ the sequence $\left(\varphi^{S_n}\right)_{n \in \mathbb{N}}$ of the evaluation of φ in the symmetric groups is vanishing. Show that Γ is sofic.

Hint Fix $\varepsilon > 0$ and a finite subset $F = \{g_1, \ldots, g_n\}$ of Γ. Write an existential sentence φ witnessing the existence of elements $g_1, \ldots, g_n$ with the multiplication rules given by Γ. Infer from the fact that φ holds in Γ that φ approximately holds in S_n for n large enough. Use this to construct an (F, ε)-approximate morphism $\Phi : \Gamma \to S_n$. ■

It is easy to deduce from Exercise 2.7.2 the following characterization of sofic groups, showing in particular that soficity is an elementary property.

Proposition 2.7.3 *If Γ is a group, the following statements are equivalent*

1. *Γ is sofic;*
2. *if φ is an existential sentence that holds in Γ, then $\lim_{n \to +\infty} \varphi^{S_n} = 0$;*
3. *if φ is a universal sentence such that $\lim_{n \to +\infty} \varphi^{S_n} = 0$, then $\varphi^{\Gamma} = 0$.*

All the results of this section carry over to the case of hyperlinear group, when one replaces the permutation groups with the unitary groups (see [111]). The analogue of Exercise 2.7.1 for the unitary groups can be proved in a similar way, considering for $u \in U_n$ and $N = kn + r$ the unitary matrix

$$\begin{pmatrix} u \otimes I_k & 0 \\ 0 & 1_{U_r} \end{pmatrix} \in U_N.$$

2.8 The Kervaire-Laudenbach Conjecture

Let Γ be a countable discrete group and $\gamma_1, \ldots, \gamma_l \in \Gamma$. Denote $w(x)$ the monomial

$$w(x) = x^{n_1} \gamma_1 \ldots x^{n_l} \gamma_l,$$

where $n_i \in \mathbb{Z}$ for $i = 1, 2, \ldots, l$. Consider the following problem: Determine if the equation

$$w(x) = 1$$

has a solution in some group extending Γ. The answer in general is "no". Consider for example the equation

$$xax^{-1}b^{-1} = 1$$

If a and b have different orders then clearly this equation has no solution in any group extending Γ. Assuming that the sum $\sum_{i=1}^{l} n_i$ of the exponents of x in $w(x)$ is nonzero is a way to rule out this obstruction. A conjecture attributed to Kervaire and Laudenbach asserts that this is enough to guarantee the existence of a solution of the equation $w(x) = 1$ in some group extending Γ.

We will show in this section that the Kervaire-Laudenbach conjecture holds for hyperlinear groups (see Definition 2.2.2). This result, first observed by Pestov in [125], will be a direct consequence of the following theorem by Gerstenhaber and Rothaus (see [69]).

Theorem 2.8.1 *Let U_n denote the group of unitary matrices of rank n. If $a_1, \ldots, a_k$ are elements of U_n then the equation*

$$x^{s_1} a_1 \cdots x^{s_k} a_k = 1$$

has a solution in U_n as long as $\sum_{i=1}^{k} s_i \neq 0$.

Proof Let $w(x, a_1, \ldots, a_k) := x^{s_1} \cdots x^{s_k} a_k$ and consider the map $f : U_n \to U_n$ defined by

$$b \mapsto w(b, a_1, \ldots, a_n).$$

We just need to prove that f is onto. Recall that U_n is a compact manifold of dimension n^2. Thus the homology group $H_{n^2}(U_n)$ is an infinite cyclic group. (A standard reference for homology theory is [67, Chap. 23].) Being continuous (and in fact smooth) f induces a homomorphism

$$f_* : H_{n^2}(U_n) \to H_{n^2}(U_n).$$

If e is a generator of $H_{n^2}(U_n)$ then

$$f_*(e) = de$$

for some $d \in \mathbb{Z}$ called the *degree* of f. In order to show that f is onto, it is enough to show that its degree is nonzero. We claim that $d = s^n$, where $s = \sum_{i=1}^{n} s_i$. Being U_n connected, the map f is homotopy equivalent to the map

$$f_s : U_n \to U_n$$

defined by

$$b \mapsto b^s.$$

By homotopy invariance of the degree of a map, f and f_s have the same degree. Therefore we just have to show that f_s has degree s^n. This follows from the facts that the generic element of U_n has s^n s-roots of unity, and that the degree of a map can be computed locally. $\square$

Theorem 2.8.1 is in fact a particular case of [69, Theorem 2], where arbitrary compact Lie groups and system of equations with possibly several variables are considered.

Let us now discuss how one can infer from Theorem 2.8.1 that the Kervaire-Laudenbach conjecture holds for hyperlinear group. Consider a word

$$w(x, y_1, \ldots, y_k) \equiv x^{s_1} y_1 x^{s_2} y_2 \cdots x^{s_k} y_k,$$

where $\sum_{i=1}^{k} s_i \neq 0$. By Theorem 2.8.1, formula

$$\sup_{y_1,\ldots,y_k} \inf_x \ell\left(w(x, y_1, \ldots, y_k)\right)$$

evaluates to 0 in any unitary group U_n. By Łoś' theorem on ultraproducts, the same formula evaluates to 0 in any ultraproduct $\prod_{\mathcal{U}} U_n$ of the unitary groups. Thus if $a_1, \ldots, a_k$ are elements of $\prod_{\mathcal{U}} U_k$

$$\inf_{x \in \prod_{\mathcal{U}} U_n} \ell\left(w(x, a_1, \ldots, a_k)\right) = 0.$$

By countable saturation of $\prod_{\mathcal{U}} U_n$ one can find $b \in \prod_{\mathcal{U}} U_n$ where the infimum is achieved. This means that

$$\ell\left(w(b, a_1, \ldots, a_k)\right) = 0$$

and hence b is a solution of the equation

$$x^{s_1} a_1 \cdots x^{s_k} a_k.$$

This shows that any ultraproduct $\prod_{\mathcal{U}} U_n$ satisfies the Kervaire-Laudenbach conjecture. Since hyperlinear groups are subgroups of $\prod_{\mathcal{U}} U_n$, they will satisfy the Kervaire-Laudenbach conjecture as well.

2.9   Other Metric Approximations and Higman's Group

The notions of sofic and hyperlinear groups as defined in Sects. 2.1 and 2.2 respectively admit natural generalizations where one considers approximate morphisms into different classes of invariant length groups. Suppose that C is a class of groups endowed with an invariant length function.

Definition 2.9.1 A countable discrete group Γ has the *C-approximation property* if for every finite subset F of $\Gamma \setminus \{1_\Gamma\}$ and every $\varepsilon > 0$ there is an (F, ε)-approximate morphism (as defined in Definition 2.1.7) from Γ endowed with the trivial length function to a group T in C, i.e. a function $\Phi : \Gamma \to T$ such that $\Phi(1_\Gamma) = 1_T$ and for every $g, h \in F$:

- $\ell_T\left(\Phi(gh)\Phi(h)^{-1}\Phi(g)^{-1}\right) < \varepsilon$;
- $\ell_T\left(\Phi(g)\right) > 1 - \varepsilon$.

The following characterization of groups locally embeddable into some class of invariant length groups can be proved with the same arguments as Sect. 2.6, where the notions of universal and existential sentence are introduced.

Proposition 2.9.2 *The following statements about a countable discrete group* Γ *(endowed with the trivial length function) are equivalent:*

1. Γ *has the* $\mathcal{C}$-*approximation property;*
2. *There is a length-preserving group homomorphism from* Γ *to an ultraproduct* $\prod_{\mathcal{U}} G_n$ *of a sequence* $(G_n)_{n \in \mathbb{N}}$ *of invariant metric groups from the class* $\mathcal{C}$;
3. *For every existential sentence* φ *in the language of invariant length groups*

$$\varphi^{\Gamma} \geq \inf \left\{ \varphi^{G} : G \in \mathcal{C} \right\}$$

where φ^{G} *denotes the evaluation of* φ *in the invariant length group* G;
4. *For every universal sentence* φ *in the language of invariant length groups*

$$\varphi^{\Gamma} \leq \sup \left\{ \varphi^{G} : G \in \mathcal{C} \right\}.$$

This characterization in particular shows that the $\mathcal{C}$-approximation property is elementary in the sense of Sect. 2.6

Observe that when $\mathcal{C}$ is a class of groups endowed with the trivial length function, the $\mathcal{C}$-approximation property coincides with the notion of local embeddability into elements of $\mathcal{C}$ as defined in Sect. 2.3. It is immediate from the definition that a group is sofic if and only if it has the $\mathcal{C}$-approximation property where $\mathcal{C}$ is the class of symmetric groups endowed with the Hamming invariant length function. Analogously hyperlinearity can be seen as the $\mathcal{C}$-approximation property where $\mathcal{C}$ is the class of unitary groups endowed with the Hilbert-Schmidt invariant length functions. *Weakly sofic groups* as defined in [70] are exactly $\mathcal{C}$-approximable groups where $\mathcal{C}$ is the class of all finite groups endowed with an invariant length function. More recently Arzhantseva and Păunescu introduced in [6] the class of *linear sofic groups*, which can be regarded as $\mathcal{C}$-approximable groups where $\mathcal{C}$ is the class of general linear groups endowed with the invariant length function

$$\ell(x) = N(x - 1)$$

where N is the usual normalized rank of matrices.

Giving a complete account of the notion of local metric approximation in group theory (not to mention other areas of mathematics) would be too long and beyond the scope of this survey. It should be nonetheless mentioned that it can be found *in nuce* in the work of Malcev. More recently it has been considered by Gromov in the paper [79] that lead to the introduction of sofic groups. The notion of $\mathcal{C}$-approximation is defined implicitly by Pestov in [125, Remark 8.6] and explicitly by Thom in [144, Definition 1.6], as well as by Arzhantseva and Cherix (Quantifying metric approximations of groups, unpublished).

A more general notion of metric approximation called *asymptotic approximation* has been more recently defined by Arzhantseva for *finitely generated groups* in [5, Definition 9]: A finitely generated group Γ is asymptotically approximated by a class of invariant length groups $\mathcal{C}$ if there is a sequence $(X_n)_{n \in \mathbb{N}}$ of finite generating subsets of Γ such that for every $n \in \mathbb{N}$ there is a $\left(B_{X_n}(n), \frac{1}{n}\right)$-approximate morphism as in Definition 2.1.7 from Γ endowed with the trivial invariant length function to an element of $\mathcal{C}$, where $B_{X_n}(n)$ denotes the set of elements of Γ that have length at most n according to the word length associated with the generating set X_n. In the particular case when the finite generating set X_n does not depend on n one obtains the notion of $\mathcal{C}$-approximation as above. It is showed in [5, Theorem 11] that several important classes of finitely generated groups, including all hyperbolic groups and one-relator groups, are asymptotically residually finite, i.e. have the asymptotic $\mathcal{C}$-approximation property where $\mathcal{C}$ is the class of residually finite groups endowed with the trivial length function. In particular these groups are asymptotically sofic. (It is currently not known if these groups are in fact sofic.)

To this day no countable discrete group is known to *not* have the $\mathcal{C}$-approximation property as in Definition 2.9.1 when $\mathcal{C}$ is any of the classes of invariant length groups mentioned above. In an attempt to find an example of a countable discrete group failing to have the $\mathcal{C}$-approximation property with respect to some natural large class of invariant length groups, Thom introduced in [144] the class $\mathcal{F}_c$ of finite commutator-contractive invariant length groups, i.e. finite groups endowed with an invariant length function ℓ satisfying

$$\ell(xyx^{-1}y^{-1}) \leq 4\ell(x)\ell(y).$$

Examples of such groups, besides finite groups endowed with the trivial length function, are finite subgroups of the unitary group of a C*-algebra A endowed with the length function

$$\ell(x) = \frac{1}{2} \|1 - x\|.$$

Corollary 3.3 of [144] shows that groups with the $\mathcal{F}_c$-approximation property form a proper subclass of the class of all countable discrete groups. More precisely Higman's group H introduced by Higman in [85] does not have the $\mathcal{F}_c$-approximation property. This is one of the few currently know examples of a group failing to have the $\mathcal{C}$-approximation property for some broad class $\mathcal{C}$ of groups endowed with a (nontrivial) invariant length function.

Higman's group H is the group with generators h_i for $i \in \mathbb{Z}/4\mathbb{Z}$ subject to the cyclic relations

$$h_{i+1}h_ih_{i+1}^{-1} = h_i^2,$$

for $i \in \mathbb{Z}/4\mathbb{Z}$ where the sum $i + 1$ is calculated modulo 4. It was first considered by Higman in [85], who showed that H is an infinite group with no nontrivial finite quotients, thus providing the first example of a finitely presented group with this property. Higman's proof in fact also shows that H is not locally embeddable into finite groups. Theorem 3.2 and Corollary 3.3 from [144] strengthen Higman's result, showing that H does not even have the $\mathcal{F}_c$-approximation property.

We will now prove, following Higman's original argument from [85], that Higman's group H is nontrivial and in fact infinite. Consider for $i \in \mathbb{Z}/4\mathbb{Z}$ the group H_i generated by h_i and h_{i+1} subjected to the relation

$$h_{i+1} h_i h_{i+1}^{-1} = h_i^2,$$

where again $i + 1$ is calculated modulo 4. It is not hard to see that every element of H_i can be expressed uniquely as $h_i^n h_{i+1}^m$ for $n, m \in \mathbb{Z}$. In particular h_i and h_{i+1} respectively generate disjoint cyclic free groups in H_i. Define H_{01} to be the free product of H_0 and H_1 amalgamated over the common free cyclic subgroup generated by h_1, and H_{23} to be the free product of H_2 and H_3 amalgamated over the common free cyclic subgroup generated by h_3. By Corollary 8.11 from [116] $\{h_0, h_2\}$ generates a free subgroup of both H_{01} and H_{23}. Define H to be the free product of H_{01} and H_{23} amalgamated over the subgroup generated by $\{h_0, h_2\}$. Again by Corollary 8.11 from [116] $\{h_1, h_3\}$ generates a free subgroup of H. In particular H is an infinite group.

As mentioned before, Higman showed that the group H is not locally embeddable into finite groups. Equivalently the system $\mathcal{R}_H(x_0, x_1, x_2, x_3)$ consisting of the relations

$$x_{i+1} x_i x_{i+1}^{-1} = x_i^2,$$

for $i \in \mathbb{Z}/4\mathbb{Z}$ has no nontrivial finite models. This means that if F is any finite group and $a_i \in F$ for $i \in \mathbb{Z}/4\mathbb{Z}$ satisfy the system $\mathcal{R}_H$, i.e. satisfy

$$a_{i+1} a_i a_{i+1}^{-1} = a_i^2,$$

for $i \in \mathbb{Z}/4\mathbb{Z}$, then

$$a_0 = a_1 = a_2 = a_3 = 1_F.$$

In fact suppose by contradiction that one of the a_i's is nontrivial. Define p to be the smallest prime number dividing the order of one of the a_i's. Without loss of generality we can assume that p divides the order of a_0. From the equation

$$a_1 a_0 a_1^{-1} = a_0^2,$$

one can obtain by induction

$$a_1^n a_0 a_1^{-n} = a_0^{2^n}.$$

In particular if n is the order of a_1 we have

$$a_0^{2^n - 1} = 1_F,$$

which shows that $2^n - 1$ is a multiple of the order of a_0 and hence of p. In particular p is an odd prime and

$$2^n \equiv 1 \ (\mathrm{mod} p).$$

Thus by Fermat's little theorem $p - 1$ divides n, contradicting the assumption that p is the smallest prime dividing the order of one of the a_i's.

A similar argument from [141] shows that the system $\mathcal{R}_H$ has no nontrivial solution in $GL_n(\mathbb{C})$ for any $n \in \mathbb{N}$. In fact suppose that $a_i \in GL_n(\mathbb{C})$ for $i \in \mathbb{Z}/4\mathbb{Z}$ satisfy the system $\mathcal{R}_H$. The relation

$$a_{i+1} a_i a_{i+1}^{-1} = a_i^2$$

implies that a_i and a_i^2 are conjugate and, in particular, have the same eigenvalues. This implies that the eigenvalues of a_i are roots of unity. Considering the Jordan canonical form of a_i shows that the absolute value of the entries of a_i^n grows at most polynomially in n. Now the relation

$$a_{i+1}^n a_i a_{i+1}^{-n} = a_i^{2^n}$$

shows that also the absolute value of the entries of $a_i^{2^n}$ grows at most polynomially in n. It follows that a_i is diagonalizable and, having roots of unity as eigenvalues, a finite order element of $GL_n(\mathbb{C})$. The same argument as before now shows that the a_i's are equal to the unity of $GL_n(\mathbb{C})$.

Andreas Thom showed in [144, Corollary 3.3] that H does not have the $\mathcal{F}_c$-approximation property where $\mathcal{F}_c$ is the class of finite groups endowed with a commutator-contractive invariant length function. This is a strengthening of Higman's result that H is not locally embeddable into finite groups, since the trivial invariant length function on a group is commutator-contractive. The following proposition is the main technical result involved in the proof; see [144, Theorem 3.2].

Proposition 2.9.3 *Suppose that G is a finite commutator contractive invariant length group and ε is a positive real number smaller than $\frac{1}{176}$. If a_i for $i \in \mathbb{Z}/4\mathbb{Z}$ are elements of G satisfying the system $\mathcal{R}_H$ up to ε, then*

$$\ell(a_i) < 4\varepsilon$$

for every $i \in \mathbb{Z}/4\mathbb{Z}$.

In view of Proposition 2.9.2 in order to conclude that Higman's group H does not have the $\mathcal{F}_c$-approximation property, it is enough to show that there is a universal sentence φ such that $\varphi^G = 0$ for every invariant length group G but $\varphi^\Gamma = 1$. Write

$$\psi(x_0, x_1, x_2, x_3) \equiv \max_{i \in \mathbb{Z}/4\mathbb{Z}} \ell(x_i) - 4 \max_{i \in \mathbb{Z}/4\mathbb{Z}} \ell(x_{i+1} x_i x_{i+1}^{-1} x_i^{-2})$$

where $i + 1$ is calculated modulo 4, and define φ to be the universal sentence

$$\sup_{x_0, x_1, x_2, x_3} f(\psi)$$

where f is the function

$$x \mapsto \min\{\max\{x, 0\}, 1\}.$$

Observe that by Proposition 2.9.3 the interpretation of φ in any commutator-contractive invariant length group is 0, while the interpretation of φ in Higman's group endowed with the trivial length function is 1.

2.10   Rank Rings and Kaplansky's Direct Finiteness Conjecture

Recall that Kaplansky's direct finiteness conjecture for a countable discrete group Γ asserts that if K is a field, then the group algebra $K\Gamma$ is directly finite. This means that if a, b are elements of $K\Gamma$ such that $ab = 1$ then also $ba = 1$. This conjecture has been confirmed when Γ is a sofic group by Elek and Szabó in [54]. An alternative proof of this result has been obtained in [30, Corollary 1.4] using the theory of cellular automata; see also [31, Sect. 8.15]. The proof of this result can be naturally presented within the framework of rank rings.

Definition 2.10.1 Suppose that R is a ring. A function $N : R \to [0, 1]$ is a *rank function* if:

- $N(1) = 1$;
- $N(x) = 0$ iff $x = 0$;
- $N(xy) \leq \min\{N(x), N(y)\}$;
- $N(x + y) \leq N(x) + N(y)$.

If N is a rank function on R then

$$d(x, y) = N(x - y)$$

defines a metric that makes the function $x \mapsto x + a$ isometric and the functions $x \mapsto xa$ and $x \mapsto ax$ contractive for every $a \in R$. A ring endowed with a rank function is called a *rank ring*.

In the context of rank rings, a *term* in the variables $x_1, \ldots, x_n$ is just a polynomial in the indeterminates $x_1, \ldots, x_n$. A *basic formula* is an expression of the form

$$N(p(x_1, \ldots, x_n))$$

where $p(x_1, \ldots, x_n)$ is a term in the variables $x_1, \ldots, x_n$. *Formulae, sentences* and their *interpretation* in a rank ring can then be defined starting from terms analogously as in the case of invariant length groups. Also the notion of elementary property, and axiomatizable class, saturated and countably saturated structure carry over without change.

Suppose that $(R_n)_{n \in \mathbb{N}}$ is a sequence of rank rings and $\mathcal{U}$ is a free ultrafilter over $\mathbb{N}$. The ultraproduct $\prod_{\mathcal{U}} R_n$ is the quotient of the product ring $\prod_n R_n$ by the ideal

$$I_{\mathcal{U}} = \left\{ (x_n) \in \prod_n R_n \,\middle|\, \lim_{n \to \mathcal{U}} N_n(x_n) = 0 \right\}.$$

The function

$$N_{\mathcal{U}}(x_n) = \lim_{n \to \mathcal{U}} N_n(x_n)$$

induces a rank function in the quotient, making $\prod_{\mathcal{U}} R_n$ a rank ring. Then all the R_n coincide with the same rank ring R the corresponding ultraproduct will be called *ultrapower* of R. The notion of *representative sequence* of an element of an ultraproduct of rank rings is defined analogously as in the case of length groups. Łoś' theorem and countable saturation of ultraproducts can be proved in this context in a way analogous to the case of invariant length groups. In particular:

Theorem 2.10.2 (Łoś) *Suppose that $\varphi(x_1, \ldots, x_k)$ is a formula for rank rings with free variables $x_1, \ldots, x_k$, $(R_n)_{n \in \mathbb{N}}$ is a sequence of rank rings, and $\mathcal{U}$ is a free ultrafilter over $\mathbb{N}$. If $a^{(1)}, \ldots, a^{(k)}$ are elements of $\prod_n R_n$ then*

$$\varphi^{\prod_{\mathcal{U}} R_n}\left(a^{(1)}, \ldots, a^{(k)}\right) = \lim_{n \to \mathcal{U}} \varphi^{R_n}\left(a_n^{(1)}, \ldots, a_n^{(k)}\right)$$

where $a^{(i)}$ is any representative sequence of $\left(a_n^{(i)}\right)_{n \in \mathbb{N}}$ for $i = 1, 2, \ldots, k$. In particular if φ is a sentence then

$$\varphi^{\prod_{\mathcal{U}} R_n} = \lim_{n \to \mathcal{U}} \varphi^{R_n}.$$

A rank ring R such that for every $x, y \in R$

$$N(xy - 1) = N(yx - 1)$$

is called a *finite rank ring*. Clearly any finite rank ring is a directly finite ring. Moreover the class of finite rank rings is axiomatizable by the formula

$$\sup_x \sup_y |N(xy - 1) - N(yx - 1)|.$$

It follows that an ultraproduct of finite rank rings is a finite rank ring and in particular a directly finite ring. Exercise 2.10.3 shows that there is a tight connection between rank functions on rings and length functions on groups.

Exercise 2.10.3 Suppose that N is a rank function on a ring R. Define

$$\ell(x) = N(x - 1),$$

for $x \in R$. Show that ℓ is a length function on the multiplicative group $R^\times$ of invertible elements of R, which is invariant if and only if R is a finite rank ring.

A natural example of finite rank rings is given by rings of matrices over an arbitrary field K. Denote by $M_n(K)$ the ring of $n \times n$ matrices with coefficients in K Suppose that ρ is the usual matrix rank on $M_n(K)$, i.e. $\rho(x)$ for $x \in M_n(K)$ is the dimension of the range of x regarded as a linear operator on a K-vector space of dimension n. Define

$$N(x) = \frac{1}{n}\rho(x)$$

for every $x \in M_n(K)$.

Exercise 2.10.4 Prove that N is a rank function on $M_n(K)$ as in Definition 2.10.1. Show that moreover $M_n(K)$ endowed with the rank N is a finite rank ring.

Another natural example of finite rank rings comes from von Neumann algebra theory: If τ is a faithful normalized trace on a von Neumann algebra M, then

$$N_\tau(x) = \tau\left(s(x)\right),$$

where $s(x)$ is the support projection of x, is a rank function on M. Moreover M endowed with the rank function N_τ is a finite rank ring.

Fix a free ultrafilter $\mathcal{U}$ over $\mathbb{N}$. In this section the symbol $\prod_{\mathcal{U}} M_n(K)$ will denote the ultraproduct with respect to $\mathcal{U}$ of the sequence of matrix rings $M_n(K)$ regarded as rank rings. By Exercise 2.10.4 and Łoś' theorem on ultraproducts $\prod_{\mathcal{U}} M_n(K)$ is a finite rank ring (and in particular a directly finite ring).

The rest of this section is dedicated to the proof from [53] that sofic groups satisfy Kaplansky's direct finiteness conjecture. The idea is that soficity of Γ together with the algebraic embedding of S_n into $M_n(K)$ obtained by sending σ to the associated permutation matrix P_σ allows one to construct an injective *-homomorphism from the group algebra $K\Gamma$ to the ultraproduct $\prod_{\mathcal{U}} M_n(K)$. In order to do this one needs some relations between the rank of a linear combination of permutation matrices and the lengths of the associated permutations. Explicit upper and lower bounds of the former in term of the latter ones are established in Exercise 2.10.5 and Exercise 2.10.6 respectively.

In the following for $\sigma \in S_n$ denote by $P_\sigma \in M_n(K)$ the associated permutation matrix as in Sect. 2.2.

Exercise 2.10.5 Show that

$$N(P_\sigma - I) \le \ell_{S_n}(\sigma)$$

where ℓ_{S_n} is the Hamming invariant length function.

Hint Define $c(\sigma)$ the number of cycles of σ (including fixed points). Show that

$$N(P_\sigma - I) = 1 - \frac{c(\sigma)}{n}$$

by induction on the number of cycles. ∎

Exercise 2.10.6 Suppose that $\sigma_1, \dots, \sigma_k \in S_n$ and $\lambda_1, \dots, \lambda_k \in K \setminus \{0\}$. Define

$$\varepsilon = \min_{1 \le i \le k} (1 - \ell(\sigma_i)).$$

Prove that

$$N\left(\sum_{i=1}^k \lambda_i P_{\sigma_i}\right) \ge \frac{1 - \varepsilon k}{k^2}.$$

Hint Denote by $\{e_i \mid i \in n\}$ the canonical basis of K^n. Recall that S_n is assumed to act on the set $n = \{0, 1, \dots, n-1\}$. Define D to be a maximal subset of n such that for $s, t \in X$ and $1 \le i, j \le k$ such that either $s \ne t$ or $i \ne j$ one has

$$\sigma_i(s) \ne \sigma_j(t).$$

Observe that if $x \in span\{e_i \mid i \in D\}$ then

$$\sum_{i=1}^k \lambda_i P_{\sigma_i}(x) \ne 0.$$

Infer that

$$N\left(\sum_{i=1}^k \lambda_i P_{\sigma_i}\right) \ge \frac{|D|}{n}.$$

By maximality of X for every $s \in n$ there are $t \in D$ and $1 \le i, j \le k$ such that

$$s = \sigma_i^{-1} \sigma_j(t)$$

When $i, ,j$ vary between 1 and k and t varies in D the expression

$$\sigma_i^{-1} \sigma_j(t)$$

attains at most $\varepsilon nk + k^2 |D|$ values. Infer that

$$\frac{|D|}{n} \geq \frac{1 - \varepsilon k}{k^2}.$$

∎

Exercises 2.10.5 and 2.10.6 allow one to define a nontrivial morphism from the group algebra $K\left(\prod_{\mathcal{U}} S_n\right)$ to $\prod_{\mathcal{U}} M_n(K)$. This is the content of Exercise 2.10.7.

Exercise 2.10.7 Define, using Exercise 2.10.5, a ring morphism $\Psi : K\left(\prod_{\mathcal{U}} S_n\right) \to \prod_{\mathcal{U}} M_n(K)$. Prove using Exercise 2.10.6 that if $x_1, \ldots, x_k \in \prod_{\mathcal{U}} S_n$ are such that $\ell_{\mathcal{U}}(x_i) = 1$ for $i = 1, 2, \ldots, k$ and $\lambda_1, \ldots, \lambda_k \in K \setminus \{0\}$ then

$$N\left(\lambda_1 x_1 + \cdots + \lambda_k x_k\right) \geq \frac{1}{k}.$$

One can now easily prove that a sofic group Γ satisfies Kaplansky's finiteness conjecture. In fact if Γ is sofic then Γ embeds into $\prod_{\mathcal{U}} S_n$ in such a way that the length of any element in the range of $\Gamma \setminus \{1_\Gamma\}$ is 1. This induces a ring morphism from $K\Gamma$ into $K\left(\prod_{\mathcal{U}} S_n\right)$. By composing it with the ring morphism described in Exercise 2.10.7, we obtain a ring morphism from $K\Gamma$ into $\prod_{\mathcal{U}} M_n(K)$ that is one to one by the second statement in Exercise 2.10.7. This shows that $K\Gamma$ is isomorphic to a subring of the directly finite ring $\prod_{\mathcal{U}} M_n(K)$. In particular $K\Gamma$ is itself directly finite.

2.11   Logic for Tracial von Neumann Algebras

In the context of tracial von Neumann algebras a *term* $p(x_1, \ldots, x_n)$ in the variables $x_1, \ldots, x_n$ is a *noncommutative *-polynomial* in $x_1, \ldots, x_n$, i.e. a polynomial in the noncommuting variables $x_1, \ldots, x_n$ and $x_1^*, \ldots, x_n^*$. A basic formula is an expression of the form

$$\tau\left(p(x_1, \ldots, x_n)\right),$$

where $p(x_1, \ldots, x_n)$ is a noncommutative *-polynomial. General formulae can be obtained from basic formulae composing with continuous functions or taking infima and suprema over norm bounded subsets of the von Neumann algebra or of the field of scalars. More formally if $\varphi_1, \ldots, \varphi_m$ are formulae and $f : \mathbb{C}^m \to \mathbb{C}$ is a continuous function then

$$f\left(\varphi_1, \ldots, \varphi_m\right)$$

is a formula. Analogously if $\varphi(x_1, \ldots, x_n, y)$ is a formula then

$$\inf_{\|y\| \leq 1} \operatorname{Re}\left(\varphi\left(x_1, \ldots, x_n, y\right)\right)$$

and

$$\inf_{|\lambda|\leq 1} \operatorname{Re} (\varphi (x_1, \ldots, x_n, \lambda))$$

are formulae. Similarly one can replace inf with sup. The interpretation of a formula in a tracial von Neumann algebra is defined in the obvious way by recursion on the complexity. For example

$$\left(\tau(x^*x)\right)^{\frac{1}{2}}$$

is a formula usually abbreviated by $\|x\|_2$ whose interpretation in a tracial von Neumann algebra (M) is the 2-norm on M associated with the trace τ. Analogously

$$\sup_{\|x\|\leq 1} \sup_{\|y\|\leq 1} \|x - y\|_2$$

is a sentence (i.e. a formula without free variables) that holds in a tracial von Neumann algebra M iff M is abelian, while

$$\sup_{\|x\|\leq 1} \inf_{|\lambda|\leq 1} \|x - \lambda\|_2$$

is a sentence which holds in (M) iff M is one-dimensional (i.e. isomorphic to $\mathbb{C}$).

The notion of elementary property and axiomatizable class of tracial von Neumann algebras are defined as in the case of length groups or rank rings. In particular the previous examples shows that the property of being abelian and the property of being one-dimensional are elementary. Exercise 2.11.1 and Exercise 2.11.2 show that the property of being a factor and, respectively, the property of being a II_1 factor are elementary.

Recall that the *center* $Z(M)$ of a von Neumann algebra M is the set of elements that commute with any other element of M. This is a weakly closed subalgebra of M and hence it is itself a von Neumann algebra. The von Neumann algebra M is called a *factor* if its center contains only the scalar multiples of the identity.

The *unitary group* $U(M)$ of M is the multiplicative group of unitary elements of M, i.e. elements u satisfying $uu^* = u^*u = 1$. Recall that it can be seen using the Borel functional calculus [14, I.4.3] that the set of linear combinations of projections of a von Neumann algebra is dense in the σ-weak topology.

Exercise 2.11.1 Suppose that M is a von Neumann algebra endowed with a faithful trace τ. Show that M is a factor if and only if for every $x \in M$

$$\|x - \tau(x)\|_2 \leq \sup_{y \in M_1} \|xy - yx\|_2. \tag{2.2}$$

Conclude that the property of being a factor is elementary.

Hint If M is not a factor then a nontrivial projection p in $Z(M)$ violates equation (2.2). If M is a factor and x is an element of the unit ball of M consider the strong closure of the convex hull of the orbit

$$\{uxu^* : u \in U(M)\}$$

of x under the action of $U(M)$ on M by conjugation. Observe that such set endowed with the Hilbert-Schmidt norm is isometrically isomorphic to a closed subset of the Hilbert space $L^2(M, \tau)$ obtained from τ via the GNS construction [14, II.6.4]. It therefore has a unique element of minimal Hilbert-Schmidt norm x_0, which by uniqueness must commute with every element of $U(M)$. Since the convex hull of $U(M)$ is dense in the unit ball of M by the Russo-Dye theorem [14, II.3.2.17], x_0 belongs to the center of M. Moreover by normality of the trace x_0 must coincide with $\tau(x)$. Thus $\tau(x)$ can be approximated in the Hilbert-Schmidt norm by convex combinations of elements of the form uxu^*, with $u \in U(M)$. Equation (2.2) easily follows from this fact. ∎

Exercise 2.11.2 Fix an irrational number $\alpha \in (0, 1)$. Using the type classification of factors prove that a factor M is II_1 if and only there is a projection of trace α. Deduce that the property of being a II_1 factor is elementary.

Hint Recall that by Theorem 1.1.4 the trace in a II_1 factor attains on projections all the values between 0 and 1. ∎

Let us now define the ultraproduct $\prod_{\mathcal{U}} M_n$ of a sequence (M_n) of tracial von Neumann algebras with respect to a free ultrafilter $\mathcal{U}$ over $\mathbb{N}$. Define $\ell^\infty(M)$ the C*-algebra of sequences

$$(a_n)_{n \in \mathbb{N}} \in \prod_n M_n$$

such that

$$\sup_n \|a_n\| < +\infty$$

endowed with the norm

$$\|(a_n)\| = \sup_n \|a_n\|.$$

The ultraproduct $\prod_{\mathcal{U}} M_n$ is the quotient of $\ell^\infty(M)$ with respect to the norm closed ideal $I_{\mathcal{U}}$ of sequences (a) such that

$$\lim_{n \to \mathcal{U}} \tau\left(a_n^* a_n\right) = 0$$

endowed with the faithful trace

$$\tau\left((a_n) + I_{\mathcal{U}}\right) = \lim_{n \to \mathcal{U}} \tau(a_n).$$

Being the quotient of a C*-algebra by a norm closed ideal, $\prod_{\mathcal{U}} M_n$ is a C*-algebra. Exercise 2.11.3 shows that in fact $\prod_{\mathcal{U}} M_n$ is always a von Neumann algebra. When the sequence (M_n) is constantly equal to a tracial von Neumann algebra M, the ultraproduct $\prod_{\mathcal{U}} M_n$ is called *ultrapower* of M and denoted by $M^{\mathcal{U}}$. The notion of *representative sequence* of an element of an ultraproduct of tracial von Neumann algebras is defined analogously as in the case of invariant length groups.

The notions of approximately finitely satisfiable (or consistent) and realized set of formulae and countably saturated structure introduced in Definitions 2.6.2 and 2.6.3 for invariant length groups admit obvious generalization to the setting of tracial von Neumann algebra. An adaptation of the proof of Exercise 2.6.4 shows that an ultraproduct of a sequence of tracial von Neumann algebras is countably saturated. If moreover the elements of the sequence are separable, and the Continuum Hypothesis is assumed, then ultraproducts are in fact saturated (cf. the discussion after Exercise 2.6.4).

Exercise 2.11.3 Show that the unit ball of $\prod_{\mathcal{U}} M_n$ is complete with respect to the 2-norm. Infer that $\prod_{\mathcal{U}} M_n$ is a von Neumann algebra.

Hint Recall that an ultraproduct of a sequence of tracial von Neumann algebra is countably saturated. Suppose that $(x_n)_{n\in\mathbb{N}}$ is a sequence in the unit ball of $\prod_{\mathcal{U}} M_n$ which is Cauchy with respect to the 2-norm. Define for every $n \in \mathbb{N}$

$$\varepsilon_n = \sup\left\{\left\|x_i - x_j\right\|_2 : i,j \geq n\right\}.$$

Observe that $(\varepsilon_n)_{n\in\mathbb{N}}$ is a vanishing sequence. Consider for every $n \in \mathbb{N}$ the formula $\varphi_n(y)$

$$\inf_{\|y\|\leq 1} \max\left\{\|x_n - y\|_2 - \varepsilon_n, 0\right\}.$$

Argue that the set $\mathcal{X}$ of formulae containing $\varphi_n(y)$ for every $n \in \mathbb{N}$ is approximately finitely satisfiable and hence realized in $\prod_{\mathcal{U}} M_n$. A realization x of $\mathcal{X}$ is such that

$$\|x_n - x\|_2 \leq \varepsilon_n$$

for every $n \in \mathbb{N}$ and hence the sequence $(x_n)_{n\in\mathbb{N}}$ converges to x. In order to conclude that $\prod_{\mathcal{U}} M_n$ is a von Neumann algebra it is enough to observe that, by completeness of its unit ball with respect to the 2-norm and Kaplansky's density theorem (see [14, I.9.1.3]), $\prod_{\mathcal{U}} M_n$ coincides with the von Neumann algebra generated by the GNS representation of $\prod_{\mathcal{U}} M_n$ associated with the canonical trace τ of $\prod_{\mathcal{U}} M_n$. ∎

Łoś' theorem on ultraproducts also holds in this context without change.

Theorem 2.11.4 (Łoś' for Tracial von Neumann Algebras) *Suppose that $\varphi(x_1, \ldots, x_k)$ is a formula with free variables $x_1, \ldots, x_k$, $(M_n)_{n\in\mathbb{N}}$ is a sequence of tracial von Neumann algebras, and $\mathcal{U}$ is a free ultrafilter over $\mathbb{N}$. If $a^{(1)}, \ldots, a^{(k)}$*

are elements of $\prod_n M_n$ then

$$\varphi^{\prod_{\mathcal{U}} M_n}\left(a^{(1)}, \ldots, a^{(k)}\right) = \lim_{n \to \mathcal{U}} \varphi^{M_n}\left(a_n^{(1)}, \ldots, a_n^{(k)}\right),$$

where $a^{(i)}$ is any representative sequence of $\left(a_n^{(i)}\right)_{n \in \mathbb{N}}$ for $i = 1, 2, \ldots, k$. In particular if φ is a sentence then

$$\varphi^{\prod_{\mathcal{U}} M_n} = \lim_{n \to \mathcal{U}} \varphi^{M_n}.$$

In particular Łoś' theorem implies that a tracial von Neumann algebra M is elementarily equivalent to any ultrapower $M^{\mathcal{U}}$ of M. This means that if φ is any sentence, then φ has the same evaluation in M and in $M^{\mathcal{U}}$. It follows that any two ultrapowers of M are elementarily equivalent. If moreover M is separable and the Continuum Hypothesis is assumed, then any two ultrapowers of M, being saturated and elementarily equivalent, are in fact isomorphic by a standard result in model theory (see Corollary 4.14 in [60]). Conversely if the Continuum Hypothesis fails then, assuming that M is a II$_1$ factor, by Theorem 4.8 in [59] there exist nonisomorphic ultrapowers of M (and in fact $2^{2^{\aleph_0}}$ many by Proposition 8.2 of [61]).

Theorem 2.11.4 together with the fact that the property of being a II$_1$ factor is elementary (see Exercise 2.11.2) shows that an ultraproduct of II$_1$ factors is again a II$_1$ factor. Analogously an ultraproduct of factors is a factor and an ultraproduct of abelian tracial von Neumann algebras is abelian.

The trace on the von Neumann algebra naturally induces an invariant length function on $U(M)$, as shown in the following exercise.

Exercise 2.11.5 Suppose that M is a tracial von Neumann algebra. Show that the function $\ell : U(M) \to [0, 1]$ defined by

$$\ell(u) = \frac{1}{2} \|u - 1\|_2$$

is an invariant length function on $U(M)$.

In particular if M is the algebra $M_n(\mathbb{C})$ of $n \times n$ complex matrices, then $U(M)$ coincides with the group U_n of $n \times n$ unitary matrices, and the induced length function on U_n coincides with the one considered in Sect. 2.2.

Recall that in Sect. 2.2 we introduced the following relation

$$\max\left\{\|x^*x - 1\|_2, \|xx^* - 1\|_2\right\}.$$

This is a formula φ^u in the logic for tracial von Neumann algebras. We have also mentioned that the polar decomposition shows that such a formula is *stable*, i.e. every approximate solution of $\varphi^u(x) = 0$ is close to an exact solution (and the estimate is uniform over all tracial von Neumann algebras). In model theoretic jargon this means that the zero-set of the interpretation φ^u in a given von Neumann algebra M—i.e. the unitary group $U(M)$—is a *definable set* as in [13, Definition 9.16]; see also [13, Proposition 9.19]. In particular it can be inferred that the unitary group of an ultraproduct of tracial von Neumann algebras is the ultraproduct of corresponding unitary groups. This is the content of Exercise 2.11.6.

Exercise 2.11.6 Suppose that $(M_n)_{n\in\mathbb{N}}$ is a sequence of tracial von Neumann algebras, and $\mathcal{U}$ is a free ultrafilter on $\mathbb{N}$. Show that any element of the unitary group of the ultraproduct $\prod_{\mathcal{U}} M_n$ admits a representative sequence of unitary elements. Conclude that $U\left(\prod_{\mathcal{U}} M_n\right)$ is isomorphic as invariant length group to the ultraproduct $\prod_{\mathcal{U}} U(M_n)$ of the sequence of unitary groups of the M_n's endowed with the invariant length described in Exercise 2.11.5.

In particular the unitary group of $\prod_{\mathcal{U}} M_n(\mathbb{C})$ can be identified with the group $\prod_{\mathcal{U}} U_n$ introduced in Sect. 2.2.

Exercise 2.11.7 The unitary group of $\prod_{\mathcal{U}} M_n(\mathbb{C})$ contains a subset X of size continuum such that $\|u - v\|_2 = \sqrt{2}$ for every pair of distinct elements u, v of X. Deduce that the same conclusion holds for the unitary group of any ultraproduct $\prod_{\mathcal{U}} M_n$ of a sequence of II_1 factors.

Hint The first statement follows directly from Exercises 2.11.6 and 2.2.6. For the second statement observe that by Theorem 1.1.4, if $(M_n)_{n\in\mathbb{N}}$ is a sequence of II_1 factors, and $\mathcal{U}$ is a free ultrafilter on $\mathbb{N}$, then $\prod_{\mathcal{U}} M_n(\mathbb{C})$ embeds into $\prod_{\mathcal{U}} M_n$. ∎

2.12 The Algebraic Eigenvalues Conjecture

Suppose that Γ is a (countable, discrete) group. Considering the particular case of the group algebra construction for the field $\mathbb{C}$ of complex numbers as in Sect. 1.1 one obtains the complex group algebra $\mathbb{C}\Gamma$ of formal finite linear combinations

$$\lambda_1 \gamma_1 + \cdots + \lambda_k \gamma_k$$

where $\lambda_i \in \mathbb{C}$ and $\gamma_i \in \Gamma$. The group ring $\mathbb{Z}\Gamma$ is the subring of $\mathbb{C}\Gamma$ of finite linear combinations

$$n_1 \gamma_1 + \cdots + n_k \gamma_k$$

where $n_i \in \mathbb{Z}$ and $\gamma_i \in \Gamma$. The natural action of $\mathbb{C}\Gamma$ on the Hilbert space $\ell^2\Gamma$ defines an inclusion of $\mathbb{C}\Gamma$ into $B\left(\ell^2\Gamma\right)$. A conjecture due to Dodziuk, Linnell, Mathai, Schick, and Yates known as *algebraic eigenvalues conjectures* (see [49]) asserts that elements x of $\mathbb{Z}\Gamma$ regarded as linear operators on $\ell^2\Gamma$ have algebraic integers as eigenvalues. Recall that a complex number is called an *algebraic integer* if it is the root of a monic polynomial with integer coefficients. The algebraic eigenvalues conjecture has been settled for sofic groups by Andreas Thom in [142]. The proof involves the notion of ultraproduct of tracial von Neumann algebras and can be naturally presented within the framework of logic for metric structures.

The complex group algebra $\mathbb{C}\Gamma$ can be endowed with a linear involutive map $x \mapsto x^*$ such that

$$(\lambda \gamma)^* = \overline{\lambda} \gamma^{-1}.$$

Recall that the trace τ on $\mathbb{C}\Gamma$ is defined by

$$\tau\left(\sum_\gamma \lambda_\gamma \gamma\right) = \lambda_{1_\Gamma}.$$

The weak closure $L\Gamma$ of $\mathbb{C}\Gamma$ in $B\left(\ell^2\Gamma\right)$ is a von Neumann algebra containing $\mathbb{C}\Gamma$ as a *-subalgebra. The trace of $\mathbb{C}\Gamma$ admits a unique extension to a faithful normalized trace τ on $L\Gamma$. Moreover $\mathbb{C}\Gamma$ is dense in the unit ball of $L\Gamma$ with respect to the 2-norm of $L\Gamma$ defined by $\|x\|_2 = \tau(x^*x)^{\frac{1}{2}}$.

In the rest of the section the matrix algebra $M_n(\mathbb{C})$ of $n \times n$ matrices with complex coefficients is regarded as a tracial von Neumann algebra endowed with the (unique) canonical normalized trace τ_n. If $\mathcal{U}$ is an ultrafilter over $\mathbb{N}$ then $\prod_{\mathcal{U}} M_n(\mathbb{C})$ denotes the ultraproduct of $M_n(\mathbb{C})$ as tracial von Neumann algebras. (Note that this is different from the ultraproduct of $M_n(\mathbb{C})$ as rank rings.) Denote by $\prod_{\mathcal{U}} M_n(\mathbb{Z})$ the closed self-adjoint subalgebra of $\prod_{\mathcal{U}} M_n(\mathbb{C})$ consisting of elements admitting representative sequences of matrices with integer coefficients. Recall that U_n denotes the group of unitary elements of $M_n(\mathbb{C})$. If σ is a permutation over n, then the associated permutation matrix P_σ is a unitary element of $M_n(\mathbb{C})$ such that $\tau(P_\sigma) = 1 - \ell(\sigma)$. This fact can be used to solve Exercise 2.12.1. The argument is analogous to the proof that sofic groups are hyperlinear (see Exercise 2.2.4).

Exercise 2.12.1 Suppose that Γ is a sofic group and U is a nonprincipal ultrafilter over N. Show that there is a trace preserving *-homomorphism of *-algebras from $\mathbb{C}\Gamma$ to $\prod_{\mathcal{U}} M_n(\mathbb{C})$ sending $\mathbb{Z}\Gamma$ into $\prod_{\mathcal{U}} M_n(\mathbb{Z})$.

Hint Use, as at the end of Sect. 2.2, the embedding of S_n into $M_n(\mathbb{C})$ sending σ to its associated permutation matrix P_σ, recalling that $\tau(P_\sigma) = 1 - \ell(\sigma)$. ∎

Since the *-homomorphism from $\mathbb{C}\Gamma$ to $\prod_{\mathcal{U}} M_n(\mathbb{C})$ obtained in Exercise 2.12.1 is trace-preserving, it extends to an embedding of $L\Gamma$ into $\prod_{\mathcal{U}} M_n(\mathbb{C})$. This can be seen using the GNS construction associated with τ (see [14, Sect. II.6.4]), or observing that by Kaplansky's density theorem [14, I.9.1.3] the operator norm unit ball of $\mathbb{C}\Gamma$ is dense in the norm unit ball of $L\Gamma$ with respect to the σ-strong topology (which coincides with the topology induced by the Hilbert-Schmidt norm associated with τ).

Suppose now that M is a von Neumann algebra, and x is an operator in M. Considering the spectral projection $p \in M$ on Ker $(x - \lambda I)$ shows that a complex number λ is an eigenvalue of x if and only if there is a nonzero projection $p \in M$ such that $(x - \lambda)p = 0$. It follows that if $\Psi : M \to N$ is an embedding of von Neumann algebra, then x and $\Psi(x)$ have the same eigenvalues. This observation together with Exercise 2.12.1 and its following observation allow one to conclude that in order to establish the algebraic eigenvalues conjecture for sofic groups it is enough to show that the elements of $\prod_{\mathcal{U}} M_n(\mathbb{Z})$ have algebraic eigenvalues. This is proved in the rest of this section using Łoś' theorem on ultraproducts together with a characterization of algebraic integers due to Thom.

Suppose in the following that $\lambda \in \mathbb{C}$ is not an algebraic integer. Theorem 2.12.2 is a consequence of Theorem 3.5 from [142].

Theorem 2.12.2 *Suppose that ε is a positive real number, and that N is a natural number. There is a positive constant $M(\lambda, N, \varepsilon)$ depending only on λ, N, and ε with the following property: For every monic polynomial p with integer coefficients such that all the zeros of p have absolute value at most N the proportion of zeros of p at distance less than $\frac{1}{M(\lambda,N,\varepsilon)}$ from λ is at most ε.*

Corollary 2.12.3 is a direct consequence of Theorem 2.12.2.

Corollary 2.12.3 *Suppose that ε is a positive real number, and that N is a natural number. Denote by $M(\lambda, N, \varepsilon)$ the positive constant given by Theorem 2.12.2. For every finite rank matrix with integer coefficients A of operator norm at most N there is a complex matrix B of the same size of operator norm at most $M(\lambda, N, \varepsilon)$ such that*

$$\|B(\lambda - A) - I\|_2 \leq \varepsilon.$$

Proof If A is a finite rank matrix with integer coefficients of norm at most N, then the minimal polynomial p_A of A is a monic polynomial with integer coefficients whose zeros have all absolute value at most N. Since λ is not an algebraic integers, λ is not an eigenvalue of p_A and hence $\lambda - A$ is invertible. Moreover by the choice of $M(\lambda, N, \varepsilon)$ the proportion of zeros of p at distance at least $\frac{1}{M(\lambda,N,\varepsilon)}$ from λ is at least $1 - \varepsilon$. This means that if p is the projection on the eigenspace corresponding to these eigenvalues, then $\tau(p) > 1 - \varepsilon$. Define $B = p(\lambda - A)^{-1}$, and observe that B has operator norm at most $M(\lambda, N, \varepsilon)$ and

$$\|B(\lambda - A) - 1\|_2 = \|1 - p\|_2 = \tau(1 - p) \leq \varepsilon.$$

$\square$

Define for every $M > 0$ the formula $\varphi_M(x)$ in the language of tracial von Neumann algebras by

$$\inf_{\|y\| \leq 1} \|My(\lambda - x) - 1\|_2.$$

By Corollary 2.12.3 if a is any element of $M_n(\mathbb{Z})$ of operator norm at most N, then

$$\varphi_{M(\lambda,N,\varepsilon)}(a) \leq \varepsilon$$

where $M(\lambda, N, \varepsilon)$ is the constant given by Theorem 2.12.2. By Łoś' theorem on ultraproducts the same is true for any element a of $\prod_{\mathcal{U}} M_n(\mathbb{Z})$ of operator norm at most N. It follows that for every $a \in \prod_{\mathcal{U}} M_n(\mathbb{Z})$ and every positive real number ε

there is $b \in \prod_{\mathcal{U}} M_n(\mathbb{C})$ such that $\|b\| \leq M(\lambda, \|a\|, \varepsilon)$ and

$$\|b(a - \lambda) - 1\|_2 \leq \varepsilon.$$

Therefore if p is a projection such that $(a - \lambda)p = 0$, then

$$\begin{aligned}
\|p\|_2 &\leq \|(b(a - \lambda) - 1)p\|_2 + \|b(a - \lambda)p\| \\
&\leq \|b(a - \lambda) - 1\|_2 \\
&\leq \varepsilon.
\end{aligned}$$

Being this true for every positive real number ε, $p = 0$. This shows that λ is not an eigenvalue of a for any element a of $\prod_{\mathcal{U}} M_n(\mathbb{Z})$, concluding the proof that the elements of $\prod_{\mathcal{U}} M_n(\mathbb{Z})$ have algebraic eigenvalues.

2.13   Entropy

2.13.1   Entropy of a Single Transformation

Suppose that X is a zero-dimensional compact metrizable space, i.e. a compact space with a countable clopen basis. Denote by T a transformation of X, i.e. a homeomorphism $T : X \to X$. A (necessarily finite) clopen partition $\mathcal{P}$ of X is *generating* for T if for every pair of distinct points x, y of X there are $C \in \mathcal{P}$ and $n \in \mathbb{N}$ such that $T^n x \in C$ and $T^n y \notin C$. Suppose that T has a generating clopen partition $\mathcal{P}$ (observe that in general this might not exist). Consider $\mathcal{P}$ as a coloring of X and $\mathcal{P}^n$ as a coloring of X^n. Denote for $n \geq 1$ by $H_n(\mathcal{P}, T)$ the number of possible colors of partial orbits of the form

$$\left(x, Tx, T^2 x, \ldots, T^n x\right) \in X^n.$$

Equivalently $H_n(\mathcal{P}, T)$ is the cardinality of the clopen partition of X consisting of sets

$$C_0 \cap T^{-1} C_1 \cap \ldots \cap T^{-n} C_n,$$

where $C_i \in \mathcal{P}$ for $i = 0, 1, 2, \ldots, n$.

A sequence $(a_n)_{n \in \mathbb{N}}$ of real numbers is called submultiplicative if $a_{n+m} \leq a_n \cdot a_m$ for every $n, m \in \mathbb{N}$.

Exercise 2.13.1 Show that the sequence

$$(H_n(\mathcal{P}, T))_{n \in \mathbb{N}}$$

is submultiplicative.

Fekete's lemma from [65] asserts that if $(a_n)_{n \in \mathbb{N}}$ is a submultiplicative sequence then the sequence $\left(\frac{1}{n} \log(a_n)\right)_{n \in \mathbb{N}}$ converges to $\inf_{n \in \mathbb{N}} \frac{1}{n} \log(a_n)$.

Exercise 2.13.2 Prove Fekete's lemma.

It follows from Fekete's lemma and Exercise 2.13.1 that the sequence

$$\left(\frac{1}{n}\log\left(H_{n+m}\left(\mathcal{P},T\right)\right)\right)_{n\in\mathbb{N}}$$

has a limit $h(T)$ that is by definition the *entropy* of the transformation T. Despite being defined in terms of the partition $\mathcal{P}$, the entropy $h(T)$ of T does not in fact depend on the choice of the clopen partition of X generating for T.

Exercise 2.13.3 Show that the entropy of T does not depend from the choice of P.

Observe that $h(T)$ is at most $\log|\mathcal{P}|$ for any clopen partition $\mathcal{P}$ of X generating for T. Recall that transformations T and T' of compact Hausdorff spaces X and X' are *topologically conjugate* if there is a homeomorphism $f : X \rightarrow X'$ such that $T' \circ f = f \circ T$ for every $x \in X$. It is not difficult to verify that entropy is a topological conjugacy invariant, that is, topologically conjugate transformations have the same entropy.

2.13.2 Entropy of an Integer Bernoulli Shift

Suppose that A is a finite alphabet of cardinality k and X is the space $A^{\mathbb{Z}}$. Consider the Bernoulli shift T on X defined by

$$T(a_i)_{i\in\mathbb{Z}} = (a_{i-1})_{i\in\mathbb{Z}}.$$

For $a \in A$ consider the clopen set

$$X_a = \{(a_i)_{i\in\mathbb{Z}} : a_0 = a\}$$

and observe that $\mathcal{P} = \{X_a \mid a \in A\}$ is a clopen partition of X generating for T. It is not hard to verify that

$$H_n\left(\mathcal{P},T\right) = k^n,$$

for every $n \in \mathbb{N}$, and hence

$$h(T) = \log(k).$$

Gottschalk's conjecture for $\mathbb{Z}$ asserts that if $f : X \rightarrow X$ is a continuous injective function such that $T \circ f = f \circ T$ then f is surjective. In view of the conjugation invariance of entropy, in order to establish Gottschalk's conjecture for $\mathbb{Z}$ it is enough to show that if Y is any proper closed T-invariant subspace of X, then the entropy

$h(T|_Y)$ of the restriction of T to Y is strictly smaller then the entropy $h(T)$ of X. Suppose that Y is a proper closed T-invariant subspace of X. Observe that

$$Q = \{X_a \cap Y : a \in A\}$$

is a generating partition for $T|_Y$. It is easy to see that

$$H_n(Q, T|_Y)$$

is the number of A-words of length n that appear in elements of Y. Since Y is a proper closed T-invariant subspace of X there is some $n \in \mathbb{N}$ such that

$$H_n(Q, T|_Y) < k^n$$

and hence

$$h(T|_Y, Y) = \inf_n \frac{1}{n} \log(H_n(Q, T|_Y)) < \log(k) = h(T, X).$$

This concludes the proof of Gottschalk's conjecture for $\mathbb{Z}$.

2.13.3 Tilings on Amenable Groups

Suppose in the following that Γ is a (countable, discrete) group. If E, F are subsets of Γ, then define

- $F^{-E} = \bigcap_{\gamma \in E} F\gamma^{-1} = \{x \in \Gamma \,|\, xE \subset F\}$;
- $F^{+E} = \bigcup_{\gamma \in E} F\gamma = \{x \in \Gamma \,|\, xE \cap F \neq \varnothing\}$;
- $\partial_E F = F^{+E} \setminus F^{-E}$.

The subset F of Γ is (E, δ)-invariant for some positive real number δ if

$$\frac{|\partial_E F|}{|F|} < \delta.$$

The group Γ is *amenable* if and only if for every finite subset E of Γ and every positive real number δ there is a finite (E, δ)-invariant subset F of Γ. This is equivalent to the existence of a Følner sequence, i.e. a sequence $(F_n)_{n \in \mathbb{N}}$ of finite subsets of Γ such that

$$\lim_n \frac{|\partial_E F_n|}{|F_n|} = 0$$

for every finite subset E of Γ.

Suppose in the following that Γ is an amenable group. A function ϕ from the set $[\Gamma]^{<\aleph_0}$ of finite subsets of Γ to the set $\mathbb{R}_+$ of positive real numbers is called:

- *subadditive* if $\phi(a \cup b) \leq \phi(a) + \phi(b)$ and $\phi(\varnothing) = 0$;

- *right invariant* if $\phi(a) = \phi(a\gamma)$ for every $\gamma \in \Gamma$.

Lindenstrauss and Weiss proved in [106] using the theory of quasi-tilings for amenable groups introduced by Ornstein and Weiss in [117] that if $\phi : [\Gamma]^{<\aleph_0} \to \mathbb{R}_+$ is a right-invariant subadditive function, then the function $E \mapsto \frac{|\phi(E)|}{|E|}$ has a Følner limit $\ell(\phi)$. This means that for every $\varepsilon > 0$ there is a finite subset E of Γ and a positive real number δ such that for every (E, δ)-invariant finite subset F of Γ

$$\left| \frac{\phi(F)}{|F|} - \ell(\phi) \right| < \varepsilon.$$

If E and E' are subsets of Γ, then an (E, E')-tiling is a subset T of Γ such that the family $\{\gamma E \mid \gamma \in T\}$ is made of pairwise disjoint sets, while the family $\{\gamma E' \mid \gamma \in T\}$ is a cover of Γ.

Exercise 2.13.4 If $1 \in E$ and $E' = EE^{-1}$, then there is an (E, E')-tiling of Γ. Moreover if T is any (E, E')-tiling, then for every finite subset F of Γ

$$\frac{|T \cap F^{-E}|}{|F|} \geq \frac{1}{|E'|} - \frac{|\partial_{E'}(F)|}{|F|}.$$

Hint To show existence consider any maximal set T with the property that the family $\{T\gamma \mid \gamma \in E\}$ contains pairwise disjoint elements. For the second statement, observe that

$$\left\{ \gamma E' \mid \gamma \in T \cap F^{+E'} \right\}$$

covers F, while

$$\left(T \cap F^{+E'} \right) \setminus \left(T \cap F^{-E} \right) \subset F^{+E'} \setminus F^{-E} = \partial_{E'}(F).$$

Thus

$$\frac{|T \cap F^{-E}|}{|F|} \geq \frac{|T \cap F^{+E'}| - |\partial_{E'}(F)|}{|F|}$$

$$\geq \frac{1}{|E'|} - \frac{|\partial_{E'}(F)|}{|F|}.$$

∎

2.13.4 Entropy of Actions of Amenable Groups

Suppose that $\Gamma \curvearrowright X$ is an action of an amenable group Γ on a zero-dimensional compact space X admitting a generating clopen partition. This means that $\mathcal{P}$ is a clopen partition of X such that for every $x, y \in X$ distinct there are $\gamma \in \Gamma$ and $C \in \mathcal{P}$ such that $\gamma \cdot x \in C$ and $\gamma \cdot y \notin C$. Regard $\mathcal{P}$ as a coloring of X and, for any finite subset F of Γ, $\mathcal{P}^F$ as a coloring of X^F. Denote by $H_F(\mathcal{P}, \Gamma, X)$ the number of possible colors of partial orbits

$$(\gamma \cdot x)_{\gamma \in F} \in X^F.$$

Equivalently $H_F(\mathcal{P}, \Gamma, X)$ is the cardinality of the clopen partition of X consisting of sets of the form

$$\bigcap_{\gamma \in F} \gamma^{-1} C_\gamma$$

for $C_\gamma \in \mathcal{P}$.

Exercise 2.13.5 The function $F \mapsto \log(H_F(\mathcal{P}, \Gamma, X))$ is right-invariant and subadditive.

It follows from Exercise 2.13.5 and the Lindenstrauss-Weiss theorem on right-invariant subadditive functions that there is a positive real number $h(\Gamma, X)$, called **entropy** of the action $\Gamma \curvearrowright X$, such that for every $\varepsilon > 0$ there is a finite subset E of Γ and a positive real number δ such that if F is an (E, δ)-invariant subset of Γ then

$$\left| h(\Gamma, X) - \frac{1}{|F|} \log(H_F(\mathcal{P}, \Gamma, X)) \right| < \varepsilon.$$

As for the case of integer actions, it can be verified that the entropy $h(\Gamma, X)$ does not depend on the chosen generating clopen partition. Recall that two actions $\Gamma \curvearrowright X$ and $\Gamma \curvearrowright X'$ are topologically conjugate if there is a homeomorphism $f : X \to Y$ such that for every $\gamma \in \Gamma$

$$f(\gamma x) = \gamma f(x).$$

It is easy to verify that topologically conjugate actions have the same entropy.

2.13.5 Entropy of Bernoulli Shifts of an Amenable Group

Suppose that Γ is an amenable group, A is a finite alphabet of cardinality k, and A^Γ is the space of Γ-sequences of elements of A. The *Bernoulli action* of Γ on X is defined by

$$\gamma \left(a_h\right)_{h\in\Gamma} = \left(a_{\gamma^{-1}h}\right)_{h\in\Gamma}.$$

The clopen partition $\mathcal{P}$ of X containing for every $a \in A$ the set

$$X_a = \{(a_h)_{h\in\Gamma} \mid a_1 = a\}$$

is generating for the Bernoulli action. It is not hard to verify that for every finite subset F of Γ

$$H_F\left(\mathcal{P}, \Gamma, A^\Gamma\right) = k^{|F|}$$

and hence

$$h\left(\Gamma, A^\Gamma\right) = \log(k).$$

Gottschalk's conjecture for the group Γ asserts that if $f : A^\Gamma \to A^\Gamma$ is a continuous injective function such that $f(\gamma x) = \gamma f(x)$ for every $\gamma \in \Gamma$ and $x \in A^\Gamma$ then f is surjective. As in the case of integers, in order to establish Gottschalk's conjecture it is enough to show that if Y is any proper closed invariant subspace of X then the entropy $h(\Gamma, Y)$ of the Bernoulli action of Γ on Y is strictly smaller than $\log(k)$. The theory of tilings is useful to show that a proper Bernoulli subshift has strictly smaller entropy.

Suppose that Y is a proper Bernoulli subshift. If $F \subset \Gamma$ is finite, define Y_F to be the set of restrictions of elements of Y to F. Observe that $H_F(\Gamma, Y) = |Y_F|$. Moreover since Y is a proper subshift of X there is a finite subset E of Γ such that $1 \in E$ and Y_E is a proper subset of A^E. Define $E' = EE^{-1}$.

Exercise 2.13.6 Show that for any finite subset F of Γ

$$\frac{1}{|F|}\log|Y_F| \le \log k - \left(\frac{1}{|E'|} - \frac{|\partial_{E'}(F)|}{|F|}\right)\log\left(\frac{k^{|E|}}{k^{|E|}-1}\right).$$

Hint Pick an (E, E')-tiling T. If F is a finite subset of Γ, then define

$$T^- = T \cap F^{-E}$$

and

$$F^* = F\backslash \bigcup_{\gamma\in T^-} \gamma E.$$

Observe that

$$Y_F \subset A^{F^*} \times \prod_{g\in T^-} Y_{gE}. \tag{2.3}$$

Deduce from Eq. (2.3) and from Exercise 2.13.4 that the conclusion holds. ∎

It follows from Exercise 2.13.6 that

$$h\left(\Gamma, Y\right) \leq \log(k) - \frac{1}{|E'|}\log\left(\frac{k^{|E|}}{k^{|E|} - 1}\right) < \log(k).$$

2.13.6 Entropy of Actions of Sofic Groups

Suppose that $\Gamma \curvearrowright X$ is an action of a sofic group on a zero-dimensional space X. As before we will assume that there is a clopen partition $\mathcal{P}$ that is generating for the action, and we will regard $\mathcal{P}$ as a coloring of X and $\mathcal{P}^n$ as a coloring of X^n. If F is a finite subset of Γ and σ is a function from Γ to S_n, define

$$H_{F,\delta}\left(\sigma, \mathcal{P}, \Gamma, X\right)$$

to be the number of colors $(C_i)_{i \in n}$ in $\mathcal{P}^n$ such that there is a sequence $(x_i)_{i \in n} \in X^n$ such that for every $\gamma \in F$ and for a proportion of at least $(1 - \delta)$ indexes $i \in n$, $\gamma^{-1} x_i$ has color $C_{\sigma_\gamma^{-1}(i)}$. Suppose that Σ is a *sofic approximation sequence* of Γ, i.e. a sequence $(\sigma_n)_{n \in \mathbb{N}}$ of maps $\sigma_n : \Gamma \to S_n$ such that for every $\gamma, \gamma' \in \Gamma$

$$\lim_{n \to +\infty} d\left(\sigma_n\left(\gamma\gamma'\right), \sigma_n(\gamma)\sigma_n\left(\gamma'\right)\right) = 0$$

and

$$\lim_{n \to +\infty} d\left(\sigma_n(\gamma), 1\right) = 1,$$

for all $\gamma \neq 1_\Gamma$. Define $h_{\Sigma, F, \delta}\left(\mathcal{P}, \Gamma, X\right)$ to be

$$\limsup_{n \to +\infty} \frac{1}{n}\log\left(H_{F,\delta}\left(\sigma_n, \mathcal{P}, \Gamma, X\right)\right).$$

The *entropy* $h_\Sigma\left(\Gamma, X\right)$ of the action $\Gamma \curvearrowright X$ relative to the sofic approximation sequence Σ is the infimum of $h_{\Sigma, F, \delta}\left(\mathcal{P}, \Gamma, X\right)$ when F varies among all finite subsets of Γ and δ varies among all positive real numbers. Observe that, as before, $h_\Sigma\left(\Gamma, X\right)$ is at most $\log|\mathcal{P}|$, it does not depend on the generating finite clopen partition chosen, and it is invariant by topological conjugation. It is shown in Sect. 5 of [100] using the so called Rokhlin lemma for sofic approximations of countable amenable groups (see Sect. 4 of [100]) that the sofic entropy associated with any sofic approximation sequence of an amenable group coincide with the classical notion of entropy for actions of amenable groups. Nonetheless the entropy of an action of a nonamenable sofic group can in general depend on the choice of the sofic approximation sequence.

2.13.7   Bernoulli Actions of Sofic Groups

Suppose that Γ is a sofic group, A is a finite alphabet, and $\Gamma \curvearrowright A^\Gamma$ is the Bernoulli action of Γ with alphabet A of cardinality k. The entropy $h_\Sigma \left(\Gamma, A^\Gamma \right)$ of the Bernoulli action with respect to any sofic approximation sequence Σ is $\log(k)$. In fact consider as before the clopen partition $\mathcal{P}$ of A^Γ consisting of the sets

$$X_a = \left\{ \left(a_\gamma \right)_{\gamma \in \Gamma} \mid a_{1_\Gamma} = a \right\}$$

for $a \in A$.

In the following we will say that $\sigma : \Gamma \to S_n$ is a *good enough sofic approximation* if

$$d \left(\sigma_{\gamma\eta}, \sigma_\gamma \sigma_\eta \right) < \varepsilon$$

for every $\eta, \gamma \in F$, where F is a large enough finite subset of Γ and ε is a small enough positive real number. It is not difficult to see that if σ is a good enough sofic approximation then

$$H_{F,\delta} \left(\mathcal{P}, \Gamma, A^\Gamma \right) = k^n.$$

It follows that

$$h_\Sigma \left(\Gamma, A^\Gamma \right) = \log(k).$$

As in the case of amenable groups, Gottschalk's conjecture for sofic groups can be proved by showing that a proper subshift of the Bernoulli shift has entropy strictly smaller than $\log(k)$. Suppose thus that Y is a proper closed invariant subspace of A^Γ. It is easy to see that if some element of A does not appear as a digit in any element of Y then $h_\Sigma \left(\Gamma, Y \right) \leq \log \left(k - 1 \right) < \log(k)$. Thus without loss of generality we can assume that all elements of A appear as digits in some element of Y. Since Y is a proper closed subset of X there is a finite subset F of Γ such that the set Y_F of restrictions of elements of Y to F is a proper subset of A^F. We will prove that, if N is the cardinality of F, then

$$\inf_{\delta > 0} h_{\Sigma, F, \delta} \left(\mathcal{P}, \Gamma, Y \right) \leq \log(k) - \frac{1}{N^2} \log \left(\frac{k^N}{k^N - 1} \right).$$

Fix an element $\left(b_\gamma \right)_{\gamma \in F}$ of $A^F \setminus Y_F$, a function $\sigma : \Gamma \to S_n$ for some $n \in \mathbb{N}$, and $\eta, \delta > 0$ such that $\delta \left| F \right| < \eta < \frac{1}{2|F|^2 + 1}$.

Lemma 2.13.7 *Suppose that* $(c_i)_{i \in n} \in A^n$. *If there is* $(x_i)_{i \in n} \in Y^n$ *such that for every* $\gamma \in F$

$$\gamma^{-1} x_i \in X_{c_{\sigma_\gamma^{-1}(i)}}$$

for a proportion of $i \in n$ larger than $(1 - \delta)$ then

$$\frac{1}{n} \left| \bigcap_{\gamma \in F} \sigma_\gamma \left[\{ i \in n : c_i = b_\gamma \} \right] \right| < \delta |F|.$$

Proof Define

$$B = \bigcap_{\gamma \in F} \sigma_\gamma \left[\{ j \in n : c_j = b_\gamma \} \right]$$

and suppose by contradiction that

$$\frac{|B|}{n} \geq \delta |F|.$$

Observe that there is a subset C of n such that

$$\frac{1}{n} |C| > 1 - \delta |F|$$

and for every $i \in C$ and $\gamma \in F$

$$\gamma^{-1} x_i \in X_{c_{\sigma_\gamma^{-1}(i)}}.$$

It follows that there is $i \in C \cap B$. Define $y = x_i$ and observe that for every $\gamma \in F$

$$y \in \bigcap_{\gamma \in F} \gamma X_{\sigma_\gamma^{-1}(i)}$$

where

$$c_{\sigma_\gamma^{-1}(i)} = b_\gamma.$$

This contradicts the fact that $(b_\gamma)_{\gamma \in F} \notin Y_F$.   □

Denote by Z the set of $i \in n$ such that for every distinct $\gamma, \gamma' \in F$ one has

$$\sigma_\gamma(i) \neq \sigma_\gamma(i).$$

Assuming that σ is a good enough sofic approximation we have

$$\frac{1}{n} |Z| > 1 - \eta$$

For every $i \in Z$ consider the set

$$V_i = \left\{ \sigma_\gamma^{-1}(i) : \gamma \in F \right\}$$

and observe that $|V_i| = F$. Take a maximal subset Z' of Z subject to the condition that V_i and V_j are disjoint for distinct i and j in Z'. Then by maximality

$$Z \subset \bigcup_{\gamma, \gamma' \in F} \sigma_\gamma^{-1} \sigma_{\gamma'} [Z']$$

and hence

$$\frac{1}{n} |Z'| \geq \frac{|Z|}{n |F|^2} \geq \frac{1 - \eta}{|F|^2}$$

Denote by S the set of choices of colors $c = (c_i)_{i \in n} \in A^n$ for which there is some $Z'' \subset Z'$ such that

$$\frac{1}{n} |Z''| > \eta$$

and for every $\gamma \in F$ and $i \in Z''$ one has that $c_{\sigma_\gamma^{-1}(i)} = b_\gamma$. For any such c one has that

$$\bigcap_{\gamma \in F} \sigma_\gamma^{-1} \left(\{ j \in n : c_j = b_\gamma \} \right) \supset Z''$$

and hence

$$\frac{1}{n} \left| \bigcap_{\gamma \in F} \sigma_\gamma^{-1} \left(\{ j \in n : c_j = b_\gamma \} \right) \right| \geq \frac{1}{n} |Z''| > \eta.$$

By Lemma 2.13.7 this shows that when σ is a good enough sofic approximation

$$H(Y, F, \delta, \sigma) \leq |A^n \setminus S|.$$

Observe that if $c = (c_i)_{i \in n} \in A^n \setminus S$ then for every $Z'' \subset Z'$ such that $|Z''| > \eta n$ there is $i \in Z''$ such that the sequences $(b_\gamma)_{\gamma \in F}$ and $\left(c_{\sigma_\gamma^{-1}(i)} \right)_{\gamma \in F}$ are distinct. This implies that there is $W \subset Z'$ such that $|W| = |Z'| - \lfloor \eta n \rfloor$ and for every $i \in W$ the sequence $\left(\sigma_\gamma^{-1}(i) \right)_{\gamma \in F}$ differs from the sequence $(b_\gamma)_{\gamma \in F}$. Therefore the number of elements of $A^n \setminus S$ is bounded from above by

$$\binom{|Z'|}{|Z'| - \lfloor \eta n \rfloor} \left(|A|^{|F|} - 1\right)^{|Z'| - \lfloor \eta n \rfloor} |A|^{n - (|Z'| - \lfloor \eta n \rfloor)|F|}. \tag{2.4}$$

Define the function $\xi(t) = -t\log(t)$ for $t \in [0, 1]$, and observe that ξ is a concave function. By Stirling's approximation formula (see [132] for a very short proof) the expression (2.4) is in turn bounded from above by

$$C\exp\left(|Z'|\,\xi\left(1 - \frac{\eta n}{|Z'|}\right) + |Z'|\,\xi\left(\frac{\eta n}{|Z'|}\right)\right) |A|^n \left(\frac{|A|^{|F|}}{|A|^{|F|} - 1}\right)^{-(|Z'| - \eta n)} \tag{2.5}$$

for some constant C not depending on $|Z'|$ or n. From the fact that ξ is a concave function and

$$\frac{1}{n}|Z'| \geq \frac{(1 - \eta)}{|F|^2} > 2\eta$$

we obtain the estimate

$$\xi\left(1 - \frac{\eta n}{|Z'|}\right) + \xi\left(\frac{\eta n}{|Z'|}\right) \leq \xi\left(1 - \frac{\eta |F|^2 n}{1 - \eta}\right) + \xi\left(\frac{\eta |F|^2 n}{1 - \eta}\, n\right).$$

It follows that the quantity (2.5) is smaller than or equal to

$$C\exp\left(n\xi\left(1 - \frac{\eta |F|^2}{1 - \eta}\right) + n\xi\left(\frac{\eta |F|^2}{1 - \eta}\right)\right) |A|^n \left(\frac{|A|^{|F|}}{|A|^{|F|} - 1}\right)^{-\left(\frac{1 - \eta}{|F|^2} - \eta\right)n}.$$

Thus

$$\frac{1}{n}\log\left(H_{F,\delta}\left(\sigma, \mathcal{P}, \Gamma, Y\right)\right)$$

$$\leq \xi\left(1 - \frac{\eta |F|^2}{1 - \eta}\right) + \xi\left(\frac{\eta |F|^2}{1 - \eta}\right)$$

$$+ \log|A| - \left(\frac{1 - \eta}{|F|^2} - \eta\right)\log\left(\frac{|A|^{|F|} - 1}{|A|^{|F|}}\right) + o(1).$$

Since this is true for every good enough sofic approximation σ, if Σ is any sofic approximation sequence then

$$h_{\Sigma,\delta,F}(\mathcal{P},\Gamma,Y)$$

$$= \limsup_{n\to+\infty} \frac{1}{n}\log\left(H_{F,\delta}(\sigma_n,\mathcal{P},\Gamma,Y)\right)$$

$$\leq \xi\left(1 - \frac{\eta|F|^2}{1-\eta}\right) + \xi\left(\frac{\eta|F|^2}{1-\eta}\right) + \log|A| - \left(\frac{1-\eta}{|F|^2} - \eta\right)\log\left(\frac{|A|^{|F|}-1}{|A|^{|F|}}\right).$$

Being this true for every $\delta, \eta < 0$ such that $\delta|F| < \eta < \frac{1}{2|F|^2+1}$, it follows that

$$\inf_{\delta>0} h_{\Sigma,\delta,F}(\mathcal{P},\Gamma,Y) \leq \log|A| - \frac{1}{|F|^2}\log\left(\frac{|A|^{|F|}-1}{|A|^{|F|}}\right)$$

as desired.

Chapter 3
Connes' Embedding Conjecture

Valerio Capraro

3.1 The Hyperfinite II$_1$ Factor

Recall that a von Neumann algebra, as defined in Sect. 1.1, is a weakly closed *-subalgebra of the algebra $B(H)$ of bounded linear operators on a Hilbert space H. A factor is just a von Neumann algebra whose center consists only of scalar multiples of the identity. An infinite-dimensional factor endowed with a (necessarily unique and weakly continuous) faithful normalized trace is a II$_1$ factor.

Definition 3.1.1 Suppose that M and N are two factors. If F is a subset of M and ε is a positive real number, then an (F, ε)-*approximate morphism* from M to N is a function $\Phi : M \to N$ such that $\Phi(1) = 1$ and for every $x, y \in F$:

- $\|\Phi(x + y) - (\Phi(x) + \Phi(y))\|_2 < \varepsilon$;
- $\|\Phi(xy) - \Phi(x)\Phi(y)\|_2 < \varepsilon$;
- $|\tau_M(x) - \tau_N(\Phi(x))| < \varepsilon$.

A II$_1$ factor M satisfies Connes' embedding conjecture (or CEC for short) if for every finite subset F of M and every positive real number ε there is a natural number n and an (F, ε)-approximate morphism $\Phi : M \to M_n(\mathbb{C})$.

Definition 3.1.2 A finite von Neumann algebra is called *hyperfinite* if it contains an increasing chain of copies of matrix algebras whose union is weakly dense.

Exercise 3.1.3 Show that a separable hyperfinite II$_1$ factor satisfies the CEC.

Hyperfiniteness is a much stronger property than satisfying the CEC. In fact it is a cornerstone result of Murray and von Neumann from [115] that there is a

V. Capraro (✉)
Center for Mathematics and Computer Science (CWI), Amsterdam, The Netherlands
e-mail: caprarovalerio@gmail.com

© Springer International Publishing Switzerland 2015
V. Capraro, M. Lupini, *Introduction to Sofic and Hyperlinear Groups and Connes' Embedding Conjecture*, Lecture Notes in Mathematics 2136,
DOI 10.1007/978-3-319-19333-5_3

unique separable hyperfinite II_1 factor up to isomorphism, usually denoted by $\mathcal{R}$. The separable hyperfinite II_1 factor admits several different characterizations. It can be seen as the group von Neumann algebra (as defined in Sect. 1.1) of the group S_∞^{fin} of finitely supported permutations of $\mathbb{N}$. Alternatively it can be described as the von Neumann algebra tensor product $\overline{\bigotimes}_n M_2(\mathbb{C})$ of countably many copies of the algebra of 2×2 matrices. This description enlightens the useful property that $\mathcal{R}$ is tensorially self-absorbing, i.e. $\mathcal{R} \simeq \mathcal{R} \overline{\otimes} \mathcal{R}$. A deep result of Connes from [43] asserts that $\mathcal{R}$ is also the unique II_1 factor that embeds in any other separable II_1 factor.

In the following *all* II_1 *factors are assumed to be separable*, apart from ultrapowers of separable II_1 factors that are never separable by Exercise 2.11.7. Moreover $\mathcal{R}$ will denote the (unique up to isomorphism) hyperfinite separable II_1 factor.

The CEC can be equivalently reformulated in terms of local representability into $\mathcal{R}$.

Definition 3.1.4 A II_1 factor M is *locally representable* in a II_1 factor N if for every finite subset F of M and for every positive real number ε there is an (F, ε)-approximate morphism from M to N.

Exercise 3.1.5 Show that a separable II_1 factor satisfies the CEC if and only if it is locally representable into $\mathcal{R}$.

Local representability can be equivalently reformulated in terms of embedding into an ultrapower, or in terms of values of universal sentences. (The notion of formula and sentence for tracial von Neumann algebras has been defined in Sect. 2.11.) The arguments are analogous to the ones seen in Sect. 2.6 and are left to the reader as Exercise 3.1.6. As in Sect. 2.6 a formula is called *universal* if it is of the form

$$\sup_{x_1} \sup_{x_2} \ldots \sup_{x_n} \psi(x_1, \ldots, x_n)$$

where no sup or inf appear in ψ.

Exercise 3.1.6 Suppose that N and M are a separable II_1 factors. Show that the following statements are equivalent:

1. M is locally representable in N;
2. M embeds into some or, equivalently, every ultrapower of N;
3. $\varphi^M \leq \varphi^N$ for every universal sentence φ.

The *universal theory* of a II_1 factor M is the function associating with any universal sentence its value in M. The *existential theory* of M is defined similarly, where universal sentences are replaced by existential sentences. Since $\mathcal{R}$ embeds in any other II_1 factor, the universal theory of $\mathcal{R}$ is *minimal* among the universal theory of II_1 factors, i.e. $\varphi^{\mathcal{R}} \leq \varphi^M$ for any II_1 factor M. In view of this observation and

Exercise 3.1.5, the particular instance of Exercise 3.1.6 when N is the hyperfinite II_1 factor implies that the following statements are equivalent:

1. M satisfies the CEC;
2. M embeds into some or, equivalently, every ultrapower of $\mathcal{R}$;
3. M has the same universal theory as $\mathcal{R}$.

In the terminology of [58], a II_1 factor N is called *locally universal* if every II_1 factor is finitely representable in N. Thus CEC asserts that the separable hyperfinite II_1 is locally universal.. The existence of a locally universal II_1 factor, which can be regarded as a sort of weak resolution of the CEC, is established in [58, Example 6.4]. It should be noted that on the other hand by [119, Theorem 2] there is no universal *separable* II_1 factor, i.e. there is no separable II_1 factor containing any other separable II_1 factor as a subfactor.

3.2 Hyperlinear Groups and Rădulescu's Theorem

Hyperlinear groups were defined in Sect. 2.2 in terms of local approximations by unitary groups of matrix algebras (see Definition 2.2.2). They can also be equivalently characterized in terms of local approximations by the unitary group of the hyperfinite II_1 factor. This is discussed in Exercise 3.2.1.

Exercise 3.2.1 Suppose that Γ is a countable discrete group. Regard Γ as an invariant length group with respect to the trivial length function, and the unitary group $U(\mathcal{R})$ of $\mathcal{R}$ as an invariant length group with respect to the invariant length described in Exercise 2.11.5. Then the following statements are equivalent:

1. Γ is hyperlinear;
2. for every finite subset F of Γ and every positive real number ε there is an (F, ε)-approximate morphism (as in Definition 2.1.7) from Γ to $U(\mathcal{R})$;
3. there is an injective group homomorphism from Γ to $U(\mathcal{R})^{\mathcal{U}}$ for some or, equivalently, any free ultrafilter $\mathcal{U}$ over $\mathbb{N}$;
4. there is a length preserving group homomorphism from Γ to $U(\mathcal{R})^{\mathcal{U}}$ for some or, equivalently, any free ultrafilter $\mathcal{U}$ over $\mathbb{N}$.

Hint The equivalence of 1 and 2 can be can be shown using the characterization of hyperlinear groups given by Proposition 2.2.9. The equivalence of 1, 3, and 4 can be established as in Exercises 2.2.3 and 2.2.10. ∎

Observe that, by Exercise 2.11.6, the ultrapower $U(\mathcal{R})^{\mathcal{U}}$ of the unitary group of $\mathcal{R}$ with respect to the free ultrafilter $\mathcal{U}$ can be identified with the unitary group $U(\mathcal{R})$ of the corresponding ultrapower of $\mathcal{R}$. Therefore, in view of Exercise 3.2.1, $U(\mathcal{R})$ can be regarded as a *universal hyperlinear group*, analogously as the ultraproduct $\prod_{\mathcal{U}} U_n$ of the finite rank unitary groups as we have seen in Sect. 2.2. Proposition 2.6.6 allows one to infer (cf. Corollary 2.6.7) that if the Continuum

Hypothesis holds then the unitary group of any ultrapower of $\mathcal{R}$ has $2^{\aleph_1}$ outer automorphisms.

It is a consequence of condition (4) in Exercise 3.2.1 that if Γ is a *countable* discrete hyperlinear group, then the group von Neumann algebra $L\Gamma$ of Γ (see Definition 1.1.5) satisfies the CEC (this a result of Rădulescu from [131]). In fact a length preserving homomorphism from Γ to $U(\mathcal{R})$ extends by linearity to a trace preserving embedding of $\mathbb{C}\Gamma$ to $\mathcal{R}^{\mathcal{U}}$. This in turn induces an embedding of $L\Gamma$ into $\mathcal{R}^{\mathcal{U}}$, witnessing that $L\Gamma$ satisfies the CEC.

With a little extra care one can show that any (not necessarily countable) subgroup Γ of the unitary group of some ultrapower of $\mathcal{R}$ has the property that the group von Neumann algebra $L\Gamma$ of Γ embeds into a (possibly different) ultrapower of $\mathcal{R}$. This is the content of Theorem 3.2.2, established by the first-named author and Păunescu in [24].

Theorem 3.2.2 *For any group Γ, the following conditions are equivalent:*

1. *Γ admits a group monomorphism into $U(\mathcal{R})$ for some free ultrafilter $\mathcal{U}$ on $\mathbb{N}$;*
2. *The group von Neumann algebra $L\Gamma$ embeds into $\mathcal{R}^{\mathcal{V}}$ for some (possibly different) free ultrafilter $\mathcal{V}$ on $\mathbb{N}$.*

If the Continuum Hypothesis holds then, as discussed in Sect. 2.11, all the ultrapowers of $\mathcal{R}$ as isomorphic. In particular one can always pick the same ultrafilter. It is not clear that this is still possible under the failure of the Continuum Hypothesis (see [21]).

In the rest of this section we will present the proof of Theorem 3.2.2 that involves the notion of product of ultrafilters. Let us denote for $k \in \mathbb{N}$ and $B \subset \mathbb{N}$ the vertical section $\{n \in \mathbb{N} : (k, n) \in B\}$ of B corresponding to k by B_k.

Definition 3.2.3 Suppose that $\mathcal{U}$, $\mathcal{V}$ are free ultrafilters on $\mathbb{N}$. The *(Fubini) product* $\mathcal{U} \times \mathcal{V}$ is the free ultrafilter on $\mathbb{N} \times \mathbb{N}$ such that $B \in \mathcal{U} \times \mathcal{V}$ if and only if the set of $k \in \mathbb{N}$ such that $B_k \in \mathcal{V}$ belongs to $\mathcal{U}$.

Exercise 3.2.4 Show that the operation $\times$ is not commutative.

Exercise 3.2.5 Suppose that $\mathcal{U}$, $\mathcal{V}$ are free ultrafilters on $\mathbb{N}$. Show that $\mathcal{U} \times \mathcal{V}$ is a free ultrafilter on $\mathbb{N} \times \mathbb{N}$. Moreover if $(a_{n,m})_{n,m \in \mathbb{N}}$ is a double-indexed sequence in $\mathbb{R}$, then

$$\lim_{n \to \mathcal{U}} \lim_{m \to \mathcal{V}} a_{n,m} = \lim_{(n,m) \to \mathcal{U} \times \mathcal{V}} a_{n,m}.$$

It follows from Exercise 3.2.5 that an *iterated ultrapower* can be regarded as a single ultrapower. More generally:

Proposition 3.2.6 *If $(M_{n,m}, \tau_{n,m})$ is a double-indexed sequence of tracial von Neumann algebras, then*

$$\prod_{\mathcal{U}} \left(\prod_{\mathcal{V}} M_{n,m} \right) \simeq \prod_{\mathcal{U} \times \mathcal{V}} M_{n,m}.$$

In particular

$$\left(M^\nu\right)^\mathcal{U} \simeq M^{\mathcal{U} \times \nu}.$$

Exercise 3.2.7 This exercise describes an amplification trick to be used in the proof of Theorem 3.2.2.

1. Let $z \in C$ such that $|z| \leq 1$ and $|z + 1| = 2$. Prove that $z = 1$.
2. Let u be a unitary in a II_1 factor with trace 1. Show that $u = 1$.
3. Let u be a unitary in a II_1 factor N and let $M_2(N)$ be the factor of two-by-two matrices with entries in N and denote by τ its normalized trace. Consider the element

$$u' = \begin{pmatrix} u & 0 \\ 0 & 1 \end{pmatrix}.$$

Show that $|\tau(u')| < 1$.

Proof of Theorem 3.2.2 (2) implies (1) is trivial. Conversely, let Γ be a group and $\theta : \Gamma \to U(\mathcal{R}^\mathcal{U})$ a group monomorphism. Define the new group homomorphism $\theta_1 : \Gamma \to U(M_2(\mathcal{R}^\mathcal{U}))$ by

$$\theta'(\gamma) = \begin{pmatrix} \theta(\gamma) & 0 \\ 0 & 1_\Gamma \end{pmatrix}.$$

By Exercise 3.2.7, this monomorphism verifies the property that $|\tau(\theta'(\gamma))| < 1$, for all $\gamma \neq 1$. Recall that $\mathcal{R}$ is up to isomorphism, the unique (separable) hyperfinite II_1 factor (see Definition 3.1.2). In particular $\mathcal{R}$ is isomorphic to $\mathcal{R} \bar{\otimes} \mathcal{R}$, as well as to the algebra $M_2(\mathcal{R})$ of 2×2 matrices over $\mathcal{R}$. As a consequence one obtains an isomorphism between $\mathcal{R}^\mathcal{U}$, $(\mathcal{R} \bar{\otimes} \mathcal{R})^\mathcal{U}$, and $M_2(\mathcal{R})$. This allows one to regard θ_1 as a group monomorphism from Γ to the unitary group of $\mathcal{R}^\mathcal{U}$ itself.

Now define θ_2 to be the map from Γ to the unitary group of $(\mathcal{R})^\mathcal{U}$ defined in the following way: If $\theta_1(\gamma)$ has a representative sequence of unitaries $(u_n)_{n \in \mathbb{N}} \in U(\mathcal{R})$ (this exists by Exercise 2.11.6) then $\theta_2(\gamma)$ has representative sequence

$$(u_n \otimes u_n)_{n \in \mathbb{N}}.$$

Identifying $(\mathcal{R} \bar{\otimes} \mathcal{R})^\mathcal{U}$ with $\mathcal{R}^\mathcal{U}$ one can regard θ_2 as a map from Γ to the unitary group of $\mathcal{R}^\mathcal{U}$. Define analogously $\theta_n : \Gamma \to U(\mathcal{R})$ for every natural number n taking the tensor product of θ_1 by itself (coordinatewise). Observe that θ_n is a monomorphism that moreover has the property that

$$|\tau(\theta_n(\gamma))| = |\tau(\theta_1(\gamma))|^n.$$

for every $\gamma \in \Gamma$. Next define θ_∞ from Γ the unitary group of the iterated ultrapower $(\mathcal{R}^{\mathcal{U}})^{\mathcal{U}}$ (which by Exercise 3.2.5 is isomorphic to $\mathcal{R}^{\mathcal{U} \times \mathcal{U}}$) assigning γ to the element of $(\mathcal{R}^{\mathcal{U}})^{\mathcal{U}}$ having

$$(\theta_n(\gamma))_{n \in \mathbb{N}}$$

as representative sequence. It is easy to check that θ_∞ is a group monomorphism verifying the additional property that $\tau(\theta(\gamma)) = 0$ for every $\gamma \neq 1_\Gamma$. It is easy to infer from this, as in the discussion after Exercise 3.2.1, that there is a trace-preserving embedding of $L\Gamma$ into $(\mathcal{R}^{\mathcal{U}})^{\mathcal{U}} \simeq \mathcal{R}^{\mathcal{U} \times \mathcal{U}}$.   □

3.3   Kirchberg's Theorem

As mentioned in the Introduction, Connes' embedding conjecture can be reformulated in several different ways. The purpose of this section is to present probably the most unexpected of such reformulations, originally proved by Kirchberg in [101].

Theorem 3.3.1 *The following statements are equivalent:*

1. *Connes' embedding conjecture holds true,*
2. $C^*(\mathbb{F}_\infty) \otimes_{\max} C^*(\mathbb{F}_\infty) = C^*(\mathbb{F}_\infty) \otimes_{\min} C^*(\mathbb{F}_\infty)$.

We refer the reader to Appendix A for all definitions needed to understand the statement of Kirchberg's theorem, as the minimal and maximal tensor product of C*-algebras and the full C*-algebra associated to a locally compact group.

We are not going to present the original proof, but a more recent and completely different one provided by Haagerup and Winsløw [82, 83]. This is fundamentally a topological proof relying on the use of the Effros-Marechal topology on the space of von Neumann algebras. To present a completely self-contained proof of this theorem is pretty much impossible, since it relies on the Tomita-Takesaki modular theory. However, we will present the idea in quite many details, leaving out a few technical lemmas, whose proof is highly non-trivial.

3.3.1   Effros-Marechal Topology on the Space of von Neumann Algebras

Let H be a Hilbert space and $vN(H)$ be the set of von Neumann algebras acting on H, that is the set of von Neumann subalgebras of $B(H)$. The Effros-Marechal topology on $vN(H)$, first introduced in Effros in [52] and Marechal in [112], can be defined is several different equivalent ways and we will in fact be playing with two different definitions.

Let us first recall some terminology. Given a net $\{C_a\}_{a \in A}$ on a direct set A, we say that a property P is *eventually* satisfied if there is $a \in A$ such that, for all $b \geq a$, C_b satisfies property P. We say that P is *frequently* satisfied if for all $a \in A$ there is $b \geq a$ such that C_b satisfies property P.

First Definition of the Effros-Marechal Topology. Let X be a compact Hausdorff space, $c(X)$ be the set of closed subsets of X and $\mathcal{N}(x)$ be the filter of neighborhoods of a point $x \in X$. $\mathcal{N}(x)$ is a direct set, ordered by inclusion. Let $\{C_a\}$ be a net of subsets of $c(X)$, define

$$\underline{\lim} C_a = \{x \in X : \forall N \in \mathcal{N}(x), N \cap C_a \neq \emptyset \text{ eventually}\}, \tag{3.1}$$

$$\overline{\lim} C_a = \{x \in X : \forall N \in \mathcal{N}(x), N \cap C_a \neq \emptyset \text{ frequently}\}. \tag{3.2}$$

It is clear that $\underline{\lim} C_\alpha \subseteq \overline{\lim} C_\alpha$, but the other inclusion is far from being true in general.

Exercise 3.3.2 Find an explicit example of a sequence C_n of closed subsets of the real interval $[0, 1]$ such that $\underline{\lim} C_n \subsetneq \overline{\lim} C_n$.

Effros proved in [51] that there is only one topology on $c(X)$, whose convergence is described by the condition:

$$C_a \to C \qquad \text{if and only if} \qquad \overline{\lim} C_a = \underline{\lim} C_a = C. \tag{3.3}$$

Exercise 3.3.3 Let M be a von Neumann subalgebra of $B(H)$. Denote by $Ball(M)$ the unit ball of M. Prove that $Ball(M)$ is weakly compact in $Ball(B(H))$.

The previous exercise allows to use Effros' convergence in our setting.

Definition 3.3.4 Let $\{M_a\} \subseteq vN(H)$ be a net. The Effros-Marechal topology is described by the following notion of convergence:

$$M_a \to M \qquad \text{if and only if} \qquad \overline{\lim} Ball(M_a) = \underline{\lim} Ball(M_a) = Ball(M). \tag{3.4}$$

Second Definition of the Effros-Marechal Topology. Recall that the strong* topology on $B(H)$ is the weakest locally convex topology making the maps

$$x \mapsto \|x\xi\| \qquad \text{and} \qquad x \mapsto \|x^*\xi\|$$

continuous for every $\xi \in H$. Let $x \in B(H)$ and let $so^*(x)$ denote the filter of neighborhoods of x with respect to the strong* topology.

Definition 3.3.5 Let $\{M_a\} \subseteq vN(H)$ be a net. We set

$$\liminf M_a = \{x \in B(H) : \forall U \in so^*(x), U \cap M_a \neq \emptyset \text{ eventually}\}. \tag{3.5}$$

Observe that $\liminf M_a$ is obviously so*-closed, contains the identity and it is closed under involution, scalar product and sum, since these operations are so*-continuous. Haagerup and Winsløw proved in [82], Lemma 2.2 and Theorem 2.3, that $\liminf M_a$ is also closed under multiplication. Therefore $\liminf M_a$ is a von Neumann algebra. Moreover, by Theorem 2.6 in [82], $\liminf M_a$ can be seen as the largest von Neumann algebra whose unit ball is contained in $\underline{\lim}Ball(M_a)$. This suggests to define $\limsup M_a$ as the smallest von Neumann algebra whose unit ball contains $\overline{\lim}Ball(M_a)$, that is clearly $(\overline{\lim}Ball(M_a))''$. Indeed, the double commutant of a subset of $B(H)$ is always a *-algebra and the double commutant theorem of von Neumann states that this is the smallest von Neumann algebra containing the set. So we are led to the following

Definition 3.3.6 Let $\{M_a\} \subseteq vN(H)$ be a net. We set

$$\limsup M_a := \left(\overline{\lim}Ball(M_a)\right)''. \tag{3.6}$$

Definition 3.3.7 The Effros-Marechal topology on $vN(H)$ is described by the following notion of convergence:

$$M_a \to M \qquad \text{if and only if} \qquad \liminf M_a = \limsup M_a = M.$$

These two definitions of the Effros-Marechal topology are shown to be equivalent in [82], Theorem 2.8.

Connes' embedding conjecture regards separable II_1 factors, that is, factors with a faithful representation into $B(H)$, with H is separable. Assuming separability on H, the Effros-Marechal topology on $vN(H)$ turns out to be separable and induced by a complete metric (i.e. $vN(H)$ is a Polish space). One possible distance is given by the Hausdorff distance between the unit balls:

$$d(M,N) = \max\left\{ \sup_{x \in Ball(M)} \left\{ \inf_{y \in Ball(N)} d(x,y) \right\}, \sup_{x \in Ball(N)} \left\{ \inf_{y \in Ball(M)} d(x,y) \right\} \right\}, \tag{3.7}$$

where d is a metric on the unit ball of $B(H)$ which induces the weak topology [112]. Since $vN(H)$ is separable, we may express its topology by using sequences instead of nets. This turns out to be particularly useful, since, if $\{M_a\} = \{M_n\}$ is a sequence, then the definition of the Effros-Marechal topology may be simplified by making use of the second definition. One has

$$\liminf M_n = \left\{x \in B(H) : \exists \{x_n\} \in \prod M_n \text{ such that } x_n \to^{s^*} x\right\}. \tag{3.8}$$

We now state the main theorem of [83]. A part from being of intrinsic interest, it allows to reformulate Kirchberg's theorem in the form that we are going to prove.

Let us fix some notation: $\mathfrak{F}_{I_{fin}}$ is the set of finite factors of *type I* acting on H, that is the set of von Neumann factors that are isomorphic to a matrix algebra. $\mathfrak{F}_I$ is the set of type I factors acting on H, that is the set of von Neumann factors that are isomorphic to some $B(H)$, with H separable. $\mathfrak{F}_{AFD}$ is the set of approximately finite dimensional or AFD factors acting on H, that is the set of factors containing an increasing chain of matrix algebras whose union is weakly dense. Finally, $\mathfrak{F}_{inj}$ is the set of injective factors acting on H, that is the set of factors that are the image of a bounded linear projection of norm 1, $P : B(H) \to M$.

Theorem 3.3.8 (Haagerup-Winsløw) *The following statements are equivalent:*

1. *$\mathfrak{F}_{I_{fin}}$ is dense in $vN(H)$,*
2. *$\mathfrak{F}_I$ is dense in $vN(H)$,*
3. *$\mathfrak{F}_{AFD}$ is dense in $vN(H)$,*
4. *$\mathfrak{F}_{inj}$ is dense in $vN(H)$,*
5. *Connes' embedding conjecture is true.*

Moreover, a separable II_1 factor M is embeddable into $\mathcal{R}^{\mathcal{U}}$ if and only if $M \in \overline{\mathfrak{F}_{inj}}$.

Since $\mathfrak{F}_{I_{fin}} \subseteq \mathfrak{F}_I \subseteq \mathfrak{F}_{AFD}$, the implications (1) $\Rightarrow$ (2) $\Rightarrow$ (3) are trivial. The implication (3) $\Rightarrow$ (1) follows from the fact that AFD factors contain, by mere definition, an increasing chain of type I_{fin} factors, whose union is weakly dense and from the first definition of the Effros-Marechal topology (Definition 3.3.4). The equivalence between (3) and (4) is a theorem by Alain Connes proved in [43]. What is really new and important in Theorem 3.3.8 is the equivalence between (4) and (5), proved in [83], Corollary 5.9, and the proof of the *last sentence*, proved in [83], Theorem 5.8.

3.3.2 Proof of Kirchberg's Theorem

By using Theorem 3.3.8, it is enough to prove the following statements:

1. If $\mathfrak{F}_{I_{fin}}$ is dense in $vN(H)$, then $C^*(\mathbb{F}_\infty) \otimes_{min} C^*(\mathbb{F}_\infty) = C^*(\mathbb{F}_\infty) \otimes_{max} C^*(\mathbb{F}_\infty)$.
2. If $C^*(\mathbb{F}_\infty) \otimes_{min} C^*(\mathbb{F}_\infty) = C^*(\mathbb{F}_\infty) \otimes_{max} C^*(\mathbb{F}_\infty)$, then $\mathfrak{F}_{inj}$ is dense in $vN(H)$.

Proof of (1) Let π be a *-representation of the algebraic tensor product $C^*(\mathbb{F}_\infty) \odot C^*(\mathbb{F}_\infty)$ into $B(H)$. Since $C^*(\mathbb{F}_\infty)$ is separable, we may assume that H is separable. In this way

$$A = \pi(C^*(\mathbb{F}_\infty) \odot \mathbb{C}1) \qquad \text{and} \qquad B = \pi(\mathbb{C}1 \odot C^*(\mathbb{F}_\infty))$$

belong to $B(H)$, with H separable. Let $\{u_n\}$ be the universal unitaries in $C^*(\mathbb{F}_\infty)$, as in Exercises A.4 and A.5. Let

$$v_n = \pi(u_n \otimes 1) \in A \qquad \text{and} \qquad w_n = \pi(1 \otimes u_n) \in B.$$

Set $M = A'' \in vN(H)$. By hypothesis, there exists a sequence $\{F_m\} \subseteq \mathfrak{F}_{\mathrm{Ifin}}$ such that $F_m \to M$ in the Effros-Marechal topology. Therefore, $A \subseteq M = \liminf F_m = \limsup F_m$. Thus, we have

$$A \subseteq \liminf F_m.$$

It follows that

$$\{v_n\} \subseteq U(A) \subseteq U(\liminf F_m) = \underline{\lim} Ball(F_m) \cap U(B(H)),$$

where the equality follows from [82], Theorem 2.6. Let $w(x)$ and $so^*(x)$ respectively the filters of weakly and strong* open neighborhoods of an element $x \in B(H)$. We have just proved that for every $n \in \mathbb{N}$ and $W \in w(v_n)$, one has $W \cap Ball(F_m) \cap U(B(H)) \neq \emptyset$ eventually in m. Let $S \in so^*(v_n)$, by [82] Lemma 2.4, there exists $W \in w(v_n)$ such that $W \cap Ball(F_m) \cap U(B(H)) \subseteq S \cap Ball(F_m) \cap U(B(H))$. Now, since the first set must be eventually non empty, also the second one must be the same. This means that we can approximate in the strong* topology v_n with elements $v_{m,n} \in U(F_m)$.

In a similar way we can find unitaries $w_{m,n}$ in F'_m such that $w_{m,n} \to^{so^*} w_n$. Indeed, we have

$$B \subseteq A' = M' = (\limsup F_m)' = \liminf F'_m,$$

where the last equality follows by the *commutant theorem* ([82], Theorem 3.5). Therefore, we may repeat word-by-word the previous argument.

Now let m be fixed, $\pi_{m,1}$ be a representation of $C^*(\mathbb{F}_\infty)$ mapping u_n to $v_{m,n}$ and $\pi_{m,2}$ be a representation of $C^*(\mathbb{F}_\infty)$ mapping u_n to $w_{m,n}$. We can find these representations because the $u'_n s$ are free and because any representation of G extends to a representation of $C^*(G)$. Notice that the ranges of these representations commute, since $v_n \in A$ and $w_n \in B$ and A, B commute. More precisely, the image of $\pi_{m,1}$ belongs into $C^*(F_m)$ and the image of $\pi_{m,2}$ belongs to $C^*(F'_m)$. So, by the universal property in Proposition A.10, there are unique representations π_m of $C^*(\mathbb{F}_\infty) \otimes_{\max} C^*(\mathbb{F}_\infty)$ such that

$$\pi_m(u_n \otimes 1) = v_{m,n} \qquad \text{and} \qquad \pi_m(1 \otimes u_n) = w_{m,n}, \qquad m, n \in \mathbb{N},$$

whose image lies into $C^*(F_m, F'_m)$.

Exercise 3.3.9 Prove that $C^*(F_m, F'_m) = F_m \otimes_{\min} F'_m$. (Hint: use Theorem 4.1.8(iii) in [91] and Lemma 11.3.11 in [92].)

By Exercise 3.3.9, $C^*(F_m, F'_m) = F_m \otimes_{\min} F'_m$ and thus π_m splits: $\pi_m = \sigma_m \otimes \rho_m$, for some σ_m, ρ_m representation of $C^*(\mathbb{F}_\infty)$ in $C^*(F_m, F'_m)$. Consequently, by the very definition of minimal C*-norm, $||\pi_m(x)|| \leq ||x||_{\min}$ for all $m \in \mathbb{N}$ and $x \in C^*(\mathbb{F}_\infty) \odot C^*(\mathbb{F}_\infty)$. Now, by Exercise A.5, the sequence $\{u_n\}$ is total and therefore π_m converges to π in the strong* point-wise sense. Namely, for all $x \in C^*(\mathbb{F}_\infty) \odot C^*(\mathbb{F}_\infty)$, one has $\pi_m(x) \to^{so^*} \pi(x)$.

Exercise 3.3.10 Prove that if $x_n \in B(H)$ converges to $x \in B(H)$ in the strong* topology, then $||x|| \leq \liminf ||x_n||$.

Therefore

$$||\pi(x)|| \leq \liminf ||\pi_m(x)|| \leq ||x||_{\min}, \qquad \forall x \in C^*(\mathbb{F}_\infty) \odot C^*(\mathbb{F}_\infty).$$

Since π is arbitrary, it follows that $||x||_{\max} \leq ||x||_{\min}$ and the proof of the first implication is complete. $\qquad \square$

In order to prove (2) we need a few more definitions and preliminary results. Given two *-representation π and ρ of the same C*-algebra A in the same $B(H)$, we say that they are unitarily equivalent, and we write $\pi \sim \rho$, if there is $u \in U(B(H))$ such that, for all $x \in A$, one has $u\pi(x)u^* = \rho(x)$. Given a family of representations π_a of the same C*-algebra in possibly different $B(H_a)$, we may define the direct sum $\bigoplus_a \pi_a$ to be a representation of A in $B(\bigoplus H_a)$ in the obvious way, that is,

$$\left(\bigoplus \pi_a \right)(x)\xi = \bigoplus (\pi_a(x)\xi_a), \qquad \text{for all } \xi = (\xi_a) \in \bigoplus H_a.$$

A family of representation is called *separating* if their direct sum is faithful (i.e. injective).

We state the following deep lemma, proved by Haagerup and Winsløw in [83], Lemma 4.3, making use of Voiculescu's Weyl-von Neumann theorem.

Lemma 3.3.11 *Let A be a unital C*-algebra and λ, ρ representations of A in $B(H)$. Assume ρ is faithful and satisfies $\rho \sim \rho \oplus \rho \oplus \cdots$. Then there exists a sequence $\{u_n\} \subseteq U(B(H))$ such that*

$$u_n \rho(x) u_n^* \to^{s^*} \lambda(x), \qquad \forall x \in A.$$

The other preliminary result is a classical theorem by Choi (see [37], Theorem 7).

Theorem 3.3.12 *Let $\mathbb{F}_2$ be the free group with two generators. Then $C^*(\mathbb{F}_2)$ has a separating family of finite dimensional representations.*

Exercise 3.3.13 Show that $\mathbb{F}_\infty$ embeds into $\mathbb{F}_2$.

Proof (Proof of (2)) By using Choi's theorem and Exercise 3.3.13 we can find a sequence σ_n of finite dimensional representations of $C^*(\mathbb{F}_\infty)$ such that $\sigma = \sigma_1 \oplus \sigma_2 \oplus \cdots$ is faithful. Replacing σ with the direct sum of countably many copies of itself, we may assume that $\sigma \sim \sigma \oplus \sigma \oplus \cdots$. Now, $\rho = \sigma \otimes \sigma$ is a faithful (by [140], IV.4.9) representation of $C^*(\mathbb{F}_\infty) \otimes_{\min} C^*(\mathbb{F}_\infty)$, because ρ splits. Moreover, ρ still satisfies $\rho \sim \rho \oplus \rho \oplus \cdots$. Now, given $M \in vN(H)$, let $\{v_n\}$, $\{w_n\}$ be strong* dense sequences of unitaries in $Ball(M)$ and $Ball(M')$, respectively. Let $\{z_n\}$ be the universal unitaries representing free generators of $\mathbb{F}_\infty$ in $C^*(\mathbb{F}_\infty)$, as in Exercises A.4 and A.5. Since the z_n's are free, we may find *-representations λ_1 and λ_2 of $C^*(\mathbb{F}_\infty)$ in $B(H)$ such that

$$\lambda_1(z_n) = v_n \qquad \text{and} \qquad \lambda_2(z_n) = w_n.$$

Since the ranges of these representations commute, we can apply the universal property in Proposition A.10 to find a representation λ of $C^*(\mathbb{F}_\infty) \otimes_{\min} C^*(\mathbb{F}_\infty)$ such that

$$\lambda(z_n \otimes 1) = v_n \qquad \text{and} \qquad \lambda(1 \otimes z_n) = w_n, \qquad \forall n \in \mathbb{N},$$

where we can use the minimal norm instead of the maximal one thank to the hypothesis of the theorem. This means that λ and ρ satisfy the hypotheses of Lemma 3.3.11 and therefore there are unitaries $u_n \in U(B(H))$ such that

$$u_n \rho(x) u_n^* \to^{so^*} \lambda(x), \qquad \forall x \in C^*(\mathbb{F}_\infty) \otimes_{\min} C^*(\mathbb{F}_\infty).$$

Define

$$M_n = u_n \rho(C^*(\mathbb{F}_\infty) \otimes \mathbb{C}1)'' u_n^*.$$

Therefore, using also Equation (3.8), we have

$$\lambda(C^*(\mathbb{F}_\infty) \otimes \mathbb{C}1) = \liminf u_n \rho(C^*(\mathbb{F}_\infty) \otimes \mathbb{C}1) u_n^* \subseteq \liminf M_n. \qquad (3.9)$$

Now, observe that

$$u_n \rho(\mathbb{C}1 \otimes C^*(\mathbb{F}_\infty)) u_n^* \subseteq M_n'$$

and therefore

$$\lambda(\mathbb{C}1 \otimes C^*(\mathbb{F}_\infty)) \subseteq \liminf M_n'. \qquad (3.10)$$

Since $\liminf M_a$ is always a von Neumann algebra, the inclusions in (3.9) and (3.10) still hold passing to the weak closure. Therefore, by (3.9), we obtain

$$M = \lambda(C^*(\mathbb{F}_\infty) \otimes \mathbb{C}1)'' \subseteq \liminf M_n, \qquad (3.11)$$

where the equality with M follows from the fact that the v_n's are strong* dense in $Ball(M)$. Analogously, by (3.10), we obtain

$$M' = \lambda(\mathbb{C}1 \otimes C^*(\mathbb{F}_\infty))'' \subseteq \liminf M_n', \qquad (3.12)$$

where the equality with M' follows from the fact that the w_n's are strong* dense in $Ball(M')$.

Now, using (3.11) and (3.12) and applying the commutant theorem ([82], Theorem 3.5), we get

$$M = M'' \supseteq \left(\liminf M'_n\right)' = \limsup(M_n)'' = \limsup M_n.$$

Therefore

$$\limsup M_n \subseteq M \subseteq \liminf M_n,$$

i.e. $M_n \to M$ in the Effros-Marechal topology. Therefore, we have proved that every von Neumann algebra M can be approximated by von Neumann algebras M_n that are constructed as strong closure of faithful representations that are direct sum of countably many finite-dimensional representations. Such von Neumann algebras are injective and therefore, we have proved that $vN_{\text{inj}}(H)$ is dense in $vN(H)$. Now, by [82], Theorem 5.2, vN_{inj} is a G_δ subset of $vN(H)$ and by [82], Theorem 3.11, and [83], Theorem 2.5, the set of all factors $\mathfrak{F}(H)$ is a dense G_δ-subset of vN(H). Since $vN(H)$ is a Polish space, we can apply Baire's theorem and conclude that also the intersection $vN_{\text{inj}}(H) \cap \mathfrak{F}(H) = \mathfrak{F}_{\text{inj}}(H)$ must be dense. □

Let $\mathcal{K}$ denote the class of groups Γ satisfying *Kirchberg's property*

$$C^*(\Gamma) \otimes_{\min} C^*(\Gamma) = C^*(\Gamma) \otimes_{\max} C^*(\Gamma)$$

The following seems to be an open and interesting problem [20].

Problem 3.3.14 Does $\mathcal{K}$ contain a countable discrete non-amenable group?

Another interesting question comes from the observation that the previous proof as well as Kirchberg's original proof uses free groups in a very strong way.

Problem 3.3.15 For which groups Γ, does the property $\Gamma \in \mathcal{K}$ implies Connes' embedding conjecture?

3.4 Connes' Embedding Conjecture and Lance's WEP

Kirchberg's theorem 3.3.1 is an astonishing link between the theory of von Neumann algebras and the theory of C*-algebras. In the same paper, Kirchberg found another profound link between these two theories: Connes' embedding conjecture is a particular case of a conjecture regarding the structure of C*-algebras, the QWEP conjecture, asking whether any C*-algebra is a quotient of a C*-algebra with Lance's WEP. It is then natural to ask whether there is a direct relation between Connes' embedding conjecture and WEP. Nate Brown answered this question in the affirmative, proving that Connes' embedding conjecture is equivalent to the analogue of Lance's WEP for separable type II_1 factors.

Definition 3.4.1 Let A, B be C*-algebras. A linear map $\phi : A \rightarrow B$ is called *positive* if $\phi(a^*a) \geq 0$, for all $a \in A$.

Definition 3.4.2 Let A, B be C*-algebras and $\phi : A \rightarrow B$ be a linear map. For every $n \in \mathbb{N}$ we can define a map $\phi_n : M_n(A) \rightarrow M_n(B)$ by setting

$$\phi_n[a_{ij}] = [\phi(a_{ij})]$$

ϕ is called *completely positive* if ϕ_n is positive for every n.

Exercise 3.4.3 Show that any *-homomorphism between C*-algebras is completely positive.

Definition 3.4.4 Let $A \subseteq B$ be two C*-algebras. We say that A is *weakly cp complemented* in B if there exists a unital completely positive map $\phi : B \rightarrow A^{**}$ such that $\phi|_A = Id_A$.

Let A be a C*-algebra. We recall that there is always a faithful representation of A into a suitable $B(H)$, for instance the GNS representation. In other words, one can always regard an abstract C*-algebra as a sub-C*-algebra of $B(H)$ through a faithful representation. Hereafter, we use the notation $A \subseteq B(H)$ to say that we have fixed one particular such representation. As we will see, the concepts we are going to introduce are independent of the representation.

Definition 3.4.5 A C*-algebra A has the *weak expectation property* (WEP, for short) if it is weakly cp complemented in $B(H)$ for a faithful representation $A \subseteq B(H)$.

The weak expectation property was introduced by Lance in [105].

Exercise 3.4.6 Show that WEP does not depend on the faithful representation $A \subseteq B(H)$.

Representation-free characterizations of WEP have been recently proven in [63, 64, 96] leading to new formulations of Connes' Embedding Problem [64].

Definition 3.4.7 A C*-algebra A has *QWEP* if it is a quotient of a C*-algebra with WEP.

QWEP conjecture states that every C*-algebra has QWEP. As mentioned above, in [101], Kirchberg proved also the following theorem.

Theorem 3.4.8 *Let M be a separable II_1 factor. The following statements are equivalent*

1. M is embeddable into some $\mathcal{R}^{\mathcal{U}}$,
2. M has QWEP.

We refer to the original paper by Kirchberg [101] and to the more recent survey by Ozawa [118] for the proof of this theorem. In these notes we focus on a technically easier but equally interesting topic: the von Neumann algebraic analogue of Lance's WEP and the proof of Brown's theorem.

Definition 3.4.9 Let $M \subseteq B(H)$ be a von Neumann algebra and $A \subseteq M$ a weakly dense C*-subalgebra. We say that M has a *weak expectation relative* to A if there exists a ucp map $\Phi : B(H) \to M$ such that $\Phi|_A = \mathrm{Id}_A$.

Theorem 3.4.10 (Brown, [16] Theorem 1.2) *For a separable type II_1 factor M the following conditions are equivalent:*

1. *M is embeddable into $\mathcal{R}^{\mathcal{U}}$.*
2. *M has a weak expectation relative to some weakly dense subalgebra.*

Observe that the notion of injectivity for von Neumann algebras can be stated also this way: $M \subseteq B(H)$ is *injective* if there exists a ucp map $\Phi : B(H) \to M$ such that $\Phi(x) = x$, for all $x \in M$. So relative weak expectation property is something just a bit less than injectivity. In fact Brown's theorem can be read by saying that weak expectation relative property is the "limit property of injectivity".

Corollary 3.4.11 *For a separable type II_1 factor the following conditions are equivalent:*

1. *M has a relative weak expectation property,*
2. *M has QWEP,*
3. *M is Effros-Marechal limit of injective factors.*

Proof It is an obvious consequence of Theorems 3.4.10 and 3.3.8. □

The equivalence between (2) and (3) has been recently generalized to every von Neumann algebra (see [2], Theorem 1.1).

We now move towards the proof of Theorem 3.4.10. Let A be a separable C*-algebra.

Definition 3.4.12 A *tracial state* on A is map $\tau : A_+ \to [0, \infty]$ such that

1. $\tau(x + y) = \tau(x) + \tau(y)$, for all $x, y \in A_+$;
2. $\tau(\lambda x) = \lambda \tau(x)$, for all $\lambda \geq 0, x \in A_+$;
3. $\tau(x^*x) = \tau(xx^*)$ for all $x \in A$;
4. $\tau(1) = 1$.

A tracial state extends to a positive functional on the whole A. We will often identify the tracial state with its extension.

Definition 3.4.13 A tracial state τ on $A \subseteq B(H)$ is called *invariant mean* if there exists a state ψ on $B(H)$ such that

1. $\psi(uTu^*) = \psi(T)$, for all $u \in U(A)$ and $T \in B(H)$;
2. $\psi|_A = \tau$.

Note 3.4.14 *Theorem 3.4.19 shows that the notion of invariant mean is independent of the choice of the faithful representation $A \subseteq B(H)$.*

In order to prove Brown's theorem we need a characterization of invariant means that will be proven in Theorem 3.4.19.

In order to prove such result we need two classical theorems and one lemma. Recall that $L^p(B(H))$ stands for the ideal of linear and bounded operators a such that $||a||_p := Tr\left((a^*a)^{\frac{p}{2}}\right)^{\frac{1}{p}}$ is finite, where Tr is the canonical semifinite trace on $B(H)$. The notation $L^p(B(H))_+$ stands for the cone of operators a in $L^p(B(H))$ that are positive, that is, $(a\xi, \xi) \geq 0$, for all $\xi \in H$. The following well known inequality is sometimes called Powers-Størmer inequality (see [128]).

Theorem 3.4.15 *Let $h, k \in L^1(B(H))_+$. Then*

$$||h - k||_2^2 \leq ||h^2 - k^2||_1,$$

In particular, if $u \in U(B(H))$ and $h \geq 0$ has finite rank, then

$$||uh^{1/2} - h^{1/2}u||_2 = ||uh^{1/2}u^* - h^{1/2}||_2 \leq ||uhu^* - h||_1^{1/2}.$$

We now use Powers-Størmer's inequality to prove a lemma. Recall that $M_q(\mathbb{C})$ stands for the von Neumann algebra of q-by-q matrices with complex entries.

Lemma 3.4.16 *Let H be a separable Hilbert space and $h \in B(H)$ be a positive, finite rank operator with rational eigenvalues and $Tr(h) = 1$. Then there exists a ucp map $\Phi : B(H) \to M_q(\mathbb{C})$ such that*

1. *$tr(\Phi(T)) = Tr(hT)$, for all $T \in B(H)$;*
2. *$|tr(\Phi(uu^*) - \Phi(u)\Phi(u^*))| < 2||uhu^* - h||_1^{1/2}$, for all $u \in U(B(H))$;*

where tr stands for the normalized trace on $M_q(\mathbb{C})$.

We prove the two statements separately.

Proof of (1) Let $v_1, \ldots, v_k \in H$ be the eigenvectors of H and $\frac{p_1}{q}, \ldots, \frac{p_k}{q}$ be the corresponding eigenvalues. Thus

1. $hv_i = \frac{p_i}{q}$;
2. $\sum_{i=1}^{k} \frac{p_i}{q} = tr(h) = 1$. It follows that $\sum p_i = q$.

Let $\{w_m\}$ be an orthonormal basis for H and let

$$B = \{v_1 \otimes w_1, \ldots, v_1 \otimes v_{p_1}\} \cup \{v_2 \otimes w_1, \ldots, v_2 \otimes w_{p_2}\} \cup \cdots \cup \{v_k \otimes w_1, \ldots v_k \otimes w_{p_k}\}$$

Let V be the subspace of $H \otimes H$ spanned by B and let $P : H \otimes H \to V$ be the orthogonal projection. Given $T \in B(H)$, observe that the following formula holds

$$Tr(P(T \otimes 1)P) = \sum_{i=1}^{k} p_i \langle Tv_i, v_i \rangle.$$

Indeed $P(T \otimes 1)P$ is representable (in the basis B) by a $q \times q$ block diagonal matrix whose blocks have dimension p_i with entries ETE, where $E : H \to \text{span}\{v_1, \ldots v_k\}$ is the orthogonal projection.

Now define $\Phi : B(H) \to M_q(\mathbb{C})$ by setting $\Phi(T) = P(T \otimes 1)P$. We have

$$tr(\Phi(T)) = \frac{1}{q}Tr(P(T \otimes 1)P)$$

$$= \sum_{i=1}^{k} \langle T\frac{p_i}{q}v_i, v_i \rangle$$

$$= \sum_{i=1}^{k} \langle Thv_i, v_i \rangle = Tr(Th).$$

The following exercise concludes the proof of the first statement.   □

Exercise 3.4.17 Prove that Φ is ucp.

Proof of (2) By writing down the matrix of $P(T \otimes 1)P(T^* \otimes 1)P$ in the basis B we have

$$Tr(P(T \otimes 1)P(T^* \otimes 1)P) = \sum_{i,j=1}^{k} |T_{i,j}|^2 \min(p_i, p_j),$$

where $T_{i,j} = \langle Tv_j, v_i \rangle$. Analogously, by writing down the matrices of $h^{1/2}T, Th^{1/2}$ and $h^{1/2}Th^{1/2}T^*$ in any orthonormal basis beginning with $\{v_1, \ldots v_k\}$ we have

$$Tr(h^{1/2}Th^{1/2}T^*) = \sum_{i,j=1}^{k} \frac{1}{q}(p_i p_j)^{1/2}|T_{i,j}|^2.$$

By using these formulae, we can make the following preliminary calculation

$$|Tr(h^{1/2}Th^{1/2}T^*) - tr(\Phi(T)\Phi(T^*))|$$

$$= \left| \sum_{i,j=1}^{k} \frac{1}{q}(p_i p_j)^{1/2}|T_{i,j}|^2 - \frac{1}{q}Tr(P(T \otimes 1)P(T^* \otimes 1)P) \right|$$

$$= \left| \sum_{i,j=1}^{k} \frac{1}{q}|T_{i,j}|^2((p_i p_j)^{1/2} - min(p_i, p_j)) \right|$$

$$\leq \sum_{i,j=1}^{k} \frac{1}{q}|T_{i,j}|^2 p_i^{1/2} \left| p_j^{1/2} - p_i^{1/2} \right|$$

$$\leq \left(\sum_{i,j=1}^{k} \frac{1}{q} |T_{i,j}|^2 p_i \right)^{1/2} \left(\sum_{i,j=1}^{k} \frac{1}{q} |T_{i,j}|^2 \left(p_i^{1/2} - p_j^{1/2} \right) \right)^{1/2}$$

$$= ||Th^{1/2}||_2 ||h^{1/2}T - Th^{1/2}||_2,$$

where the first inequality follows by the property $min(p_i, p_j) \leq p_i$ and the second one is obtained by applying the classical Hölder inequality. Now suppose that $T \in U(B(H))$, so that $||Th^{1/2}||_2 = ||h^{1/2}||_2 = 1$. Using again the Powers-Størmer inequality, we can continue the previous computation as follows:

$$||h^{1/2}T - Th^{1/2}||_2 = ||Th^{1/2}T^* - h^{1/2}||_2$$

$$\leq ||ThT^* - T||_1^{1/2}$$

We can now conclude the proof. Indeed, using the triangle inequality, the previous computation and the Cauchy-Schwarz inequality, we find

$$|Tr(\Phi(TT^*) - \Phi(T)\Phi(T^*))|$$

$$\leq |1 - Tr(h^{1/2}Th^{1/2}T^*)| + ||ThT^* - h||_1^{1/2}$$

$$= |Tr(ThT^*) - Tr(h^{1/2}Th^{1/2}T^*)| + ||ThT^* - h||_1^{1/2}$$

$$= |Tr((Th^{1/2} - h^{1/2}T)h^{1/2}T^*)| + ||ThT^* - h||_1^{1/2}$$

$$\leq ||h^{1/2}T^*||_2 ||Th^{1/2} - h^{1/2}T||_2 + ||ThT^* - h||_1^{1/2}$$

$$\leq 2||ThT - h||_1^{\frac{1}{2}},$$

where the last inequality follows by using the fact that T is unitary and by applying Powers-Størmer inequality once again.   □

The last preliminary result needed for the proof of Theorem 3.4.19 is classical theorem by Choi [36].

Theorem 3.4.18 *Let A, B be two C*-algebras and $\Phi : A \to B$ be a ucp map. Then*

$$\{a \in A : \Phi(aa^*) = \Phi(a)\Phi(a^*), \Phi(a^*a) = \Phi(a^*)\Phi(a)\}$$

$$= \{a \in A : \Phi(ab) = \Phi(a)\Phi(b), \Phi(ba) = \Phi(b)\Phi(a), \forall b \in A\}.$$

We are now ready to prove a useful characterization of invariant means, first proved in [102], Proposition 3.2.

Theorem 3.4.19 *Let τ be a tracial state on $A \subseteq B(H)$. Then the following are equivalent:*

1. τ is an invariant mean.

2. *There exists a sequence of ucp maps $\Phi_n : A \to \mathbb{M}_{k(n)}$ such that*

   (a) $\|\Phi_n(ab) - \Phi_n(a)\Phi_n(b)\|_2 \to 0$ *for all* $a, b \in A$,
   (b) $\tau(a) = \lim_{n\to\infty} tr(\Phi_n(a))$, *for all* $a \in A$.

3. *For any faithful representation $\rho : A \to B(H)$ there exists a ucp map $\Phi : B(H) \to \pi_\tau(A)''$ such that $\Phi(\rho(a)) = \pi_\tau(a)$, for all $a \in A$, where π_τ stands for the GNS representation associated to τ.*[1]

Proof of (1) $\Rightarrow$ (2) Let τ be an invariant mean with respect to the faithful representation $\rho : A \to B(H)$. Thus we can find a state ψ on $B(H)$ which extends τ and such that $\psi(uTu^*) = \psi(T)$, for all $u \in U(A)$ and for all $T \in B(H)$. Since the normal states are weak* dense in the dual of $B(H)$ and they can be represented in the form $Tr(h\cdot)$, with $h \in L^1(B(H))$, we can find a net $h_\lambda \in L^1(B(H))$ such that $Tr(h_\lambda T) \to \psi(T)$, for all $T \in B(H)$. Moreover, the representatives h_λ can be chosen to be positive and with trace 1. Now, since $\psi(uTu^*) = \psi(T)$, it follows that $Tr(uh_\lambda u^* T) = Tr(h_\lambda u^* Tu) \to \psi(u^* Tu) = \psi(T)$ and thus $Tr(h_\lambda T) - Tr((uh_\lambda u^*)T) \to 0$, for all $T \in B(H)$, i.e. $h_\lambda - uh_\lambda u^* \to 0$ in the weak topology of $L^1(B(H))$. Now let $\{U_n\}$ be an increasing family of finite sets of unitaries whose union have dense linear span in A and $\varepsilon = \frac{1}{n}$. Let $U_n = \{u_1, \ldots, u_n\}$. Fixed n, let us consider the convex hull of the set $\{u_1 h_\lambda u_1^* - h_\lambda, \ldots, u_n h_\lambda u_n^* - h_\lambda\}$. Its weak closure contains 0 (because of the previous observation) and coincide with the 1-norm closure, by the Hahn-Banach separation theorem. Thus there exists a convex combination of the h_λ's, say h, such that

1. $Tr(h) = 1$,
2. $\|uhu^* - h\|_1 < \varepsilon, \forall u \in U_n$,
3. $|Tr(uh) - \tau(u)| < \varepsilon, \forall u \in U_n$.

Moreover, since finite rank operators are norm dense in $L^1(B(H))$, we can assume without loss of generality that h has finite rank with rational eigenvalues. Now we can apply Lemma 3.4.16 in order to construct a sequence of ucp maps $\Phi_n : B(H) \to \mathbb{M}_{k(n)}$ such that

1. $Tr(\Phi_n(u)) \to \tau(u)$,
2. $|Tr(\Phi_n(uu^*)) - \Phi_n(u)\Phi_n(u^*)| \to 0$,

for every unitary in a countable set whose linear span is dense in A. So the property 2(b) is true for unitaries and consequently, being a linear property, it holds for all operators.

---

[1] We briefly recall the GNS construction (see [68, 134]). Setting $\langle x, y \rangle := \tau(x^* y)$, one obtains a possibly singular inner product on A. Let I be the subspace of elements x such that $\langle x, x \rangle = 0$ and consider the Hilbert space H obtained by completing A/I with respect to $\langle \cdot, \cdot \rangle$. Using the fact that I is a left ideal, one sees that the operator $\pi_\tau(x)$ defined by $\pi_\tau(x)(y) = xy$ passes to a linear and bounded operator from H to itself and then the mapping $x \to \pi_\tau(x)$ is a representation of A into $B(H)$. It turns out that this representation is always faithful: if A is unital, then this is immediate. Otherwise, one has to use a more subtle argument using approximate identities in C*-algebras.

In order to prove 2(a), we observe that $\Phi_n(uu^*) - \Phi_n(u)\Phi_n(u^*) \geq 0$ and thus the following inequality holds

$$||1 - \Phi_n(u)\Phi_n(u^*)||_2^2 \leq ||1 - \Phi_n(u)\Phi_n(u^*)|| tr(\Phi_n(uu^*) - \Phi_n(u)\Phi_n(u^*))$$

Since the right hand side tends to zero, also the left hand side must converge to 0. Now define $\Phi = \oplus\Phi_n : A \to \prod M_{k(n)} \subseteq \ell^\infty(\mathcal{R})$ and compose with the quotient map $p : \ell^\infty(\mathcal{R}) \to \mathcal{R}^\mathcal{U}$. The previous inequality shows that if u is a unitary such that $||\Phi_n(uu^*) - \Phi_n(u)\Phi_n(u^*)||_2 \to 0$ and $||\Phi_n(u^*u) - \Phi_n(u^*)\Phi_n(u)||_2 \to 0$, then u falls in the multiplicative domain of $p \circ \Phi$. But such unitaries have dense linear span in A and hence the whole A falls in the multiplicative domain of $p \circ \Phi$ (by Choi's Theorem 3.4.18). By definition of ultraproduct this just means that $||\Phi_n(ab) - \Phi_n(a)\Phi_n(b)||_2 \to 0$, for all $a \in A$.   □

Proof of (2) $\Rightarrow$ (3) Let $\Phi_n : A \to M_{k(n)}$ be a sequence of ucp maps with the properties stated in the theorem. By identifying each $M_{k(n)}$ with a unital subfactor of $\mathcal{R}$ we can define a ucp map $\tilde{\Phi} : A \to \ell^\infty(\mathcal{R})$ by $x \to (\Phi_n(x))_n$. Since the $\Phi_n's$ are asymptotically multiplicative in the 2-norm one gets a τ-preserving *-homomorphism $A \to \mathcal{R}^\mathcal{U}$ by composing with the quotient map $p : \ell^\infty(\mathcal{R} \to \mathcal{R}^\mathcal{U}$. Note that the weak closure of $p \circ \tilde{\Phi}(A)$ into $\mathcal{R}^\mathcal{U}$ is isomorphic to $\pi_\tau(A)''$. Thus we are in the following situation

$$A \xrightarrow{\tilde{\Phi}} \ell^\infty(\mathcal{R}) \xrightarrow{p} \mathcal{R}^\mathcal{U} \qquad \supseteq \qquad \overline{p \circ \tilde{\Phi}(A)}^w \cong \pi_\tau(A)''$$

$$A \downarrow^p$$
$$B(H)$$
$$\downarrow^i$$
$$B(K)$$

where K is a representing Hilbert space for $\ell^\infty(\mathcal{R})$ and i is a natural embedding induced by an embedding $H \to K$, whose existence is guaranteed by the fact that H is separable. Since $\ell^\infty(\mathcal{R})$ is injective, there is a projection $E : B(K) \to \ell^\infty(\mathcal{R})$ of norm 1. Let $F : \mathcal{R}^\mathcal{U} \to \pi_\tau(A)''$ be a conditional expectation (see [140], Proposition 2.36), one has

$$A \xrightarrow{\tilde{\Phi}} l^{\infty}(\mathcal{R}) \xrightarrow{p} \mathcal{R}^{\mathcal{U}} \xrightarrow{F} \overline{\pi_{\tau}(A)'' \cong p \circ \tilde{\Phi}(A)}^{w}$$

$$A \downarrow \rho$$

$$B(H)$$

$$\downarrow i \quad E$$

$$B(K)$$

Define $\Phi : B(H) \to \pi_{\tau}(A)''$ by setting $\Phi = FpEi$. One has $\Phi(\rho(a)) = \pi_{\tau}(a)$. $\qquad \square$

Proof of (3) $\Rightarrow$ (1) The hypothesis $\Phi(a) = \pi_{\tau}(a)$ guarantees that Φ is multiplicative on A, since π_{τ} is a representation. By Choi's Theorem 3.4.18 it follows that $\Phi(aTb) = \pi_{\tau}(a)\Phi(T)\pi_{\tau}(b)$, for all $a, b \in A, T \in B(H)$. Let τ'' be the vector trace on $\pi_{\tau}(A)''$ and consider $\tau'' \circ \Phi$. Clearly it extends τ. Moreover it is invariant under the action of $U(A)$, indeed

$$(\tau'' \circ \Phi)(u^* T u) = \tau''(\pi_{\tau}(u)^* \Phi(T)\pi_{\tau}(u)) = \tau''(\Phi(T)) = (\tau'' \circ \Phi)(T)$$

Hence τ is an invariant mean. $\qquad \square$

The following proposition was also proved by Nate Brown in [16].

Proposition 3.4.20 *Let M be a separable II_1 factor. There exists a *-monomorphism $\rho : C^*(\mathbb{F}_\infty) \to M$ such that $\rho(C^*(\mathbb{F}_\infty))$ is weakly dense in M.*

Proof Observe that $C^*(\mathbb{F}_\infty)$ can be viewed as inductive limit of free products of copies of itself. This can be proven by partitioning the set of generators in a sequence X_n of countable sets, by defining $A_n = C^*(X_1, \ldots, X_n)$ and by observing that $A_n = A_{n-1} * C^*(X_n) \cong C^*(\mathbb{F}_\infty)$, where $*$ stands for the free product with amalgamation over the scalars. Now, by Choi's Theorem 3.3.12 we can find a sequence of integers $\{k(n)\}_{n \in \mathbb{N}}$ and a unital *-monomorphism $\sigma : A \to \prod_{n \in \mathbb{N}} \mathbb{M}_{k(n)}$. Note that we may naturally identify each A_i with a subalgebra of A and hence, restricting σ to this copy, get an injection of A_i into $\prod \mathbb{M}_{k(n)}$.

Assume we can prove the existence of a sequence of unital *-homomorphism $\rho_i : A_i \to M$ such that:

1. Each ρ_i is injective;
2. $\rho_{i+1}|_{A_i} = \rho_i$ where we identify A_i with the "left side" of $A_i * C^*(\mathbb{F}_\infty) = A_{i+1}$;
3. The union of $\{\rho_i(A_i)\}$ is weakly dense in M.

then we would have completed the proof. Indeed, it would be enough to define ρ as the union of the ρ_i's.

The purpose is then to prove existence of such a sequence ρ_i. To this end we first choose an increasing sequence of projections of M such that $\tau_M(p_i) \to 1$. Then we define the orthogonal projections $q_n = p_n - p_{n-1}$ and consider the II_1 factors $Q_i = q_i M q_i$. Now, by the division property of II_1 factors (see Theorem 1.1.4), we can find a unital embedding $\prod \mathbb{M}_{k(n)} \to Q_i \subseteq M$. By composing with σ, we get a sequence of embeddings $A \to M$, which will be denoted by σ_i. Now $p_i M p_i$ is separable and thus there is a countable family of unitaries whose finite linear combinations are dense in the weak topology. Hence we can find a *-homomorphism $\pi_i : C^*(\mathbb{F}_\infty) \to p_i M p_i$ with weakly dense range (take the generators of $\mathbb{F}_\infty$ into $C^*(\mathbb{F}_\infty)$ and map them into that total family of unitaries). Now we define

$$\rho_1 = \pi_1 \oplus \left(\bigoplus_{j \geq 2} \sigma_j|_{A_1} \right) : A_1 \to p_1 M p_1 \oplus (\Pi_{j \geq 2} Q_j) \subseteq M$$

ρ_1 is a *-monomorphism, since each σ_i is already faithful on the whole A.

Now define a *-homomorphism $\theta_2 : A_2 = A_1 * C^*(\mathbb{F}_\infty) \to p_2 M p_2$ as the free product of the *-homomorphism $A_1 \to p_2 M p_2$ defined by $x \to p_2 \rho_1(x) p_2$ and $\pi_2 : C^*(\mathbb{F}_\infty) \to p_2 M p_2$. We then set

$$\rho_2 = \theta_2 \oplus \left(\bigoplus_{j \geq 3} \sigma_j|_{A_2} \right) : A_2 \to p_2 M p_2 \oplus (\Pi_{j \geq 3} Q_j) \subseteq M$$

Clearly $\rho_2|_{A_1} = \rho_1$. In general, we construct a map $\theta_{n+1} : A_n * C^*(\mathbb{F}_\infty) \to p_{n+1} M p_{n+1}$ as the free product of the cutdown (by p_{n+1}) of ρ_n and π_n. This map need not be injective and hence we take a direct sum with $\oplus_{j \geq n+2} \sigma_j|_{A_{n+1}}$ to remedy this deficiency. These maps have all the required properties and hence the proof is complete (note that the last property follows from the fact that the range of each θ_n is weakly dense in $p_{n+1} M p_{n+1}$). $\square$

Theorem 3.4.21 (Brown [16]) *Let M be a separable II_1 factor and $\mathcal{U}$ be a free ultrafilter on the natural numbers. The following conditions are equivalent:*

1. *M is embeddable into $\mathcal{R}^{\mathcal{U}}$.*
2. *M has the weak expectation property relative to some weakly dense subalgebra.*

Proof of (1) $\Rightarrow$ (2) Let M be embeddable into $\mathcal{R}^{\mathcal{U}}$. By Proposition 3.4.20, we may replace M with a weakly dense subalgebra A isomorphic to $C^*(\mathbb{F}_\infty)$. We want to prove that M has the weak expectation property relative to A. Let τ the unique normalized trace on M, more precisely we will prove that $\pi_\tau(M)$ has the weak expectation property relative to $\pi_\tau(A)$. Indeed τ is faithful and w-continuous and hence $\pi_\tau(M)$ and $\pi_\tau(A)$ are respectively copies of M and A and $\pi_\tau(A)$ is still weakly dense in $\pi_\tau(M)$. We first prove that $\tau|_A$ is an invariant mean. Take $\{u_n\}$ universal generators of $\mathbb{F}_\infty$ into A. Let n be fixed, since $u_n \in \mathcal{R}^{\mathcal{U}}$, then u_n is $||\cdot||_2$-ultralimit of unitaries in $\mathcal{R}$ (see Exercise 2.11.6). On the other hand, the unitary matrices are weakly dense in $U(\mathcal{R})$ and hence they are $||\cdot||_2$-dense in $U(\mathcal{R})$ (since weakly closed convex subsets coincide with the $||\cdot||_2$-closed convex ones (see, e.g., [87])). Thus we can find a sequence of unitary matrices which converges to u_n in norm $||\cdot||_2$. Let σ be the mapping which sends each u_n to such a sequence. Since the u_n's have no relations, we can extend σ to a *-homomorphism $\sigma : C^*(\mathbb{F}_\infty) \to \prod M_k(\mathbb{C}) \subseteq \ell^\infty(\mathcal{R})$. Let $p : \ell^\infty(\mathcal{R}) \to \mathcal{R}^{\mathcal{U}}$ be the quotient mapping. By the 2-norm convergence we have $(p \circ \sigma)(x) = x$ for all $x \in C^*(\mathbb{F}_\infty)$. Let $p_n : \prod_{k=1}^\infty M_k(\mathbb{C}) \to M_n(\mathbb{C})$ be the projection, by the definition of the trace in $\mathcal{R}^{\mathcal{U}}$, we have

$$\tau(x) = \lim_{n \to \mathcal{U}} tr_n(p_n(\sigma(x))),$$

where tr_n is the normalized trace on $M_n(\mathbb{C})$. Now we can apply Theorem 3.4.19(2) by setting $\phi_n = p_n \circ \sigma$ (they are ucp since they are *-homomorphisms) and conclude that $\tau|_A$ is an invariant mean. Now consider $\pi_\tau(M) \subseteq B(H)$ and $\pi_\tau(A) = \pi_{\tau|_A}(A) \subseteq B(H)$. By Theorem 3.4.19 there exists a ucp map $\Phi : B(H) \to \pi_\tau(A)'' = \pi_\tau(M)$ such that $\Phi(a) = \pi_\tau(a)$. Thus M has the weak expectation property relative to $C^*(\mathbb{F}_\infty)$. $\qquad\square$

Proof of (2) $\Rightarrow$ (1) Let $A \subseteq M \subseteq B(H)$, with A weakly dense in M, and $\Phi : B(H) \to M$ a ucp map which restricts to the identity on A. Let τ be the unique normalized trace on M. After identifying A with $\pi_\tau(A)$, we are under the hypothesis of Theorem 3.4.19(3) and thus $\tau|_A$ is an invariant mean. By Theorem 3.4.19 it follows that there exists a sequence $\phi_n : A \to \mathbb{M}_{k(n)}$ such that

1. $||\phi_n(ab) - \phi_n(a)\phi_n(b)||_2 \to 0$ for all $a, b \in A$,
2. $\tau(a) = \lim_{n \to \infty} tr_n(\phi_n(a))$, for all $a \in A$.

Let $p : \ell^\infty(\mathcal{R}) \to \mathcal{R}^{\mathcal{U}}$ be the quotient mapping. The previous properties guarantee that the ucp mapping $\Phi : A \to \mathcal{R}^{\mathcal{U}}$, defined by setting $\Phi(x) = p(\{\phi_n(x)\})$ is a *-homomorphism which preserves $\tau|_A$. It follows that Φ is injective. Indeed, $\Phi(x) = 0 \Rightarrow \Phi(x^*x) = 0 \Rightarrow \tau(x^*x) = 0 \Rightarrow x = 0$. Observe now that the weak closure of A into $\mathcal{R}^{\mathcal{U}}$ is isomorphic to M (they are algebraically isomorphic and have the same trace) and hence M embeds into $\mathcal{R}^{\mathcal{U}}$. $\qquad\square$

3.5 Algebraic Reformulation of the Conjecture

In this section we present a new line of research that has been initially designed by Rădulescu, with his proof that Connes' embedding conjecture is equivalent to a non-commutative analogue of Hilbert's 17th problem, and continued by Klep and Schweighofer first and Juschenko and Popovich afterwards, who arrived to a purely algebraic reformulation of Connes embedding conjecture. This section is merely descriptive and serves to introduce the reader to a new field of research, which, though motivated by the Connes embedding conjecture, is quite far from geometric group theory and operator theory, which are the main topics of this monography.

The reader interested in technical details of this approach, is referred to the original papers [89, 103, 130].

We begin with a short description of the original formulation of Hilbert's problem and we show hot to get to Rădulescu's formulation through a series of generalizations.

Let $\mathbb{R}[x_1, \ldots, x_n]$ denote the ring of polynomials with n indeterminates and real coefficients and $\mathbb{R}(x_1, \ldots, x_n)$ denote its quotient field. A polynomial $f \in \mathbb{R}[x_1, \ldots, x_n]$ is called *non-negative* if, for all $(x_1, \ldots, x_n) \in \mathbb{R}^n$, one has $f(x_1, \ldots, x_n) \geq 0$.

Problem 3.5.1 (Hilbert's 17th Problem) Can every non-negative polynomial be expressed as sum of squares of elements belonging to $\mathbb{R}(x_1, \ldots, x_n)$?

This problem was solved in the affirmative by Emil Artin [4], who provided an abstract proof of existence of such a sum. More recently, Delzell [48] provided an explicit algorithm.

More recently, scholars have been looking for challenging generalizations of this problem, the most intuitive of which is the one concerning matrices. Consider positive semi-definite matrices with entries in $\mathbb{R}[x_1, \ldots, x_n]$, that is, matrices that are positive semi-definite for all substitution $(x_1, \ldots, x_n)$. Observe that the matrix analogue of a square, is a positive semi-definite symmetric matrix. Indeed, every matrix B such that $B = A^*A$, is also symmetric, that is $B^* = B$; conversely, every positive semi-definite symmetric matrix B can be rooted (by functional calculus) and so it is a square: $B = (\sqrt{B})^2$.

Problem 3.5.2 Can all positive semi-definite matrices with entries in $\mathbb{R}[x_1, \ldots, x_n]$ be written as sum of squares of symmetric matrices with entries in $\mathbb{R}(x_1, \ldots, x_n)$?

This problem was solved in the affirmative independently by Gondard and Ribenoim [74] and Procesi and Schacher [129]. A constructive solution has been provided much later by Hillar and Nie [86].

In order to present the version of this problem using operator theory, we need to pass through a geometric analogue. To this end, observe that a polynomial $f \in \mathbb{R}[x_1, \ldots, x_n]$ is just a function from $\mathbb{R}^n$ to $\mathbb{R}$. So one may attempt to extend Hilbert's 17th problem from polynomial on $\mathbb{R}^n$ to more general functions defined on manifolds. Recall that an n-manifold M is called irreducible if for any embedding of the n-sphere S^{n-1} into M there exists an embedding of the n-ball B^n into M such that the image of the boundary of B^n coincides with the image of S^{n-1}.

Problem 3.5.3 (Geometric Analogue of Hilbert's 17th Problem) Let M be a paracompact irreducible analytic manifold and $f : M \to \mathbb{R}$ be a non-negative analytic function. Can f be expressed as a sum of squares of meromorphic functions?

Recall that meromorphic functions are those functions which are analytic on the whole domain expect for a set of isolated points, which are their poles. So, rational functions are meromorphic and one can thus recognize a generalization of Hilbert's 17th problem. In its generality, this problem is still open. A complete solution is known only for dimension $n = 2$ (see [28]) and for compact manifolds (see [133]).

The basic idea to get to Rădulescu's analogue in terms of operator theory is to generalize analytic functions with formal series. Let $Y_1, \ldots, Yn$ be n indeterminates. Define

$$I_n = \{(i_1, \ldots, i_p), p \in \mathbb{N}, i_1, \ldots, i_p \in \{1, \ldots, n\}\}.$$

For each $I = (i_1, \ldots, i_p) \in I_n$, set $Y_I = Y_{i_1} \cdot \ldots \cdot Y_{i_p}$ and define the space

$$V = \left\{ \sum_{I \in I_n} \alpha_I Y_I, \alpha_I \in \mathbb{C} : \forall R > 0, \left\| \sum_{I \in I_n} \alpha_I Y_I \right\|_R := \sum_I |\alpha_I| R^{|I|} < \infty \right\}.$$

Rădulescu showed in [130], Proposition 2.1, that V is a Fréchet space and so it carries a natural weak topology $\sigma(V, V^*)$ induced by its dual space. To generalize the notion of "square" and "sum of squares" to this setting, we need to introduce a notion of symmetry in V. Since V is a space of formal series, this is done in the obvious way. We set $(Y_{i_1} \cdot \ldots \cdot Y_{i_p})^* := Y_{i_p} \cdot \ldots \cdot Y_{i_1}$ and $\alpha^* := \bar{\alpha}$. This mapping can clearly be extended to an adjoint operation on V. Also, we observe that formal series are, by definition, possibly infinite sums and so there is no hope, in general, to have them expressed as a finite sum of squares. This motivates the need to use weak limits of sum of squares, instead of just the finite sums of squares.

Definition 3.5.4 We say that a formal series $q \in V$ is a sum of squares if it is in the weak closure of the set of the elements of the form p^*p, for $p \in V$.

We now observe that the original formulation of Hilbert's 17th problem concerns matrices with *real* entries and its geometric variant concerns *real* valued analytic functions. Recalling that the operator analogue of real valued functions are self-adjoint operators, it follows that the right setting in which Hilbert's 17th problem can be generalized is that of self-adjoint operators. We then introduce the space $V_{sa} = \{v \in V : v^* = v\}$. It remains only to generalize the notion of positivity.

Definition 3.5.5 A self-adjoint operator $v \in V_{sa}$ is called positive semidefinite if for every $N \in \mathbb{N}$ and for every N-tuple of self-adjoint matrices $X_1, \ldots, X_N$, one has

$$tr(p(X_1, \ldots, X_N)) \geq 0.$$

The cone of positive semi-definite operators is denoted V_{sa}^+.

One last step is needed to get to the operator analogue of Hilbert's 17th problem. Indeed, in case of polynomials, one has $Y_I - Y_{\tilde{I}} = 0$, for every permutation $\tilde{I}$ of I. Since this is no longer the case in the non-commutative world of formal series, we need to identify series which differ by a permutation.

Definition 3.5.6 Two elements $p, q \in V_{sa}^+$ are called *cyclic equivalent* if $p - q$ is weak limit of sums of scalars multiples of monomials of the form $Y_I - Y_{\tilde{I}}$, where $\tilde{I}$ is a cyclic permutation of I.

Problem 3.5.7 (Non-commutative Analogue of Hilbert's 17th Problem) Is every element of V_{sa}^+ cyclic equivalent to a weak limit of sums of squares?

As mentioned, the interest in this problem comes from the fact that it is equivalent to Connes' embedding conjecture.

Theorem 3.5.8 (Rădulescu) *The following statements are equivalent:*

1. *Connes' embedding conjecture is true;*
2. *Problem 3.5.7 has a positive answer.*

The proof of this theorem is quite involved and we refer the interested reader to the original paper by Rădulescu. Here we move on to further developments, which led to a purely algebraic reformulation of Connes' embedding conjecture. Such a purely algebraic reformulation is unexpected since all formulations we have discussed so far have a strong topological component. We discuss two different, though similar, purely algebraic reformulations of Connes' embedding conjecture; one due to Klep and Schweighofer [103], the other to Juschenko and Popovich [89].

We start by discussing Klep and Schweighofer's approach. This differs from Rădulescu in many parts, the most important of which is that they do not use formal series but they get back to polynomials. Using finite objects instead of infinite objects is the ultimate reason why they are able to get rid of every topological condition.

Let K be either the real or the complex field and let V denote the ring of polynomials on n indeterminates with coefficients in K. Instead of using Rădulescu's adjoint operation they define the adjoint operation acting identically on monomials and switching each coefficient with its conjugate. As before, the set of self-adjoint elements is denoted V_{sa}. Moreover, instead of using the cyclic equivalence, they define two polynomial p and q to be equivalent when their difference is a sum of commutators.

Definition 3.5.9 A polynomial $f \in V$ is called positive semidefinite if for every $N \in \mathbb{N}$ and for every contractions $A_1, \ldots, A_n \in \mathbb{M}_N(\mathbb{R})$ one has

$$tr(f(A_1, \ldots, A_n)) \geq 0.$$

The set of positive semidefinite elements is denoted by V^+.

The introduction of the quadratic module is the major difference with Rădulescu's approach.

Definition 3.5.10 A subset $M \subseteq V_{sa}$ is called quadratic module if the following hold:

1. $1 \in M$;
2. $M + M \subseteq M$;
3. $p^*Mp \subseteq M$, for all $p \in V$.

The quadratic module generated by the elements $1 - X_1^2, \ldots, 1 - X_n^2$ is denoted by Q.

Theorem 3.5.11 (Klep-Schweighofer) *The following statements are equivalent:*

1. *Connes' embedding conjecture is true;*
2. *For every $f \in V^+$ and for every $\varepsilon > 0$, there exists $q \in Q$ such that $f + \varepsilon$ is equivalent to q, in the sense that $f + \varepsilon - q$ is a sum of commutators.*

One may wish to be able to replace the quadratic module by the more standards squares, namely elements of the form v^*v. We conclude this section by describing the approach by Juschenko and Popovich, which is indeed aimed to this.

One way to reformulate Klep and Schweighofer's approach is by considering the free associative algebra $K(X)$ generated by a countable family of self-adjoint elements $X = (X_1, X_2 \ldots)$. Thus, a polynomial $f \in V$ is just an element of $K(X)$. Klep-Schweighofer's theorem affirms that Connes' embedding conjecture is equivalent to the statement that every positive semidefinite element of $K(X)$ cab be written as an element of the quadratic module, up to a sum of commutators and ε, with ε arbitrarily small.

Instead of considering the free associative algebra $K(X)$, Juschenko and Popovich considered the group *-algebra $\mathcal{F}$ of the countably generated free group $\mathbb{F}_\infty = \langle u_1, u_2, \ldots \rangle$. With this choice they were able to simplify Klep and Schweighofer's theorems in two ways. First, instead of using the quadratic module, they were able to use standard squares; second, instead of considering every polynomial f, they were able to consider only polynomials of degree at most two, in the variables u_i. Before presenting the theorem, we redefine the notion of positivity in this new context.

Definition 3.5.12 An element $f \in \mathcal{F}$ with n indeterminates is called positive semidefinite if for all $m \geq 1$ and all n-tuples of unitary matrices $U_1, \ldots, U_n$ of dimension m, one has

$$tr(f(U_1, \ldots, U_n)) \geq 0.$$

Theorem 3.5.13 (Juschenko-Popovich) *The following statements are equivalent:*

1. *Connes' embedding conjecture is true;*
2. *For every self-adjoint positive semidefinite $f \in \mathcal{F}$ and for every $\varepsilon > 0$, one has $f + \varepsilon = g + c$, where g is a sum of squares (elements of the for v^*v) and c is a sum of commutators.*

3.6 Brown's Invariant

Most of research about Connes' embedding conjecture has been focusing on impressive reformulations of it, that is, on finding apparently very far statements that turn out to be eventually equivalent to the original conjecture.

Over the last couple of years another point of view has been also taken, mostly due to Nate Brown's paper [17]. He assumes that a fixed separable II_1 factor M verifies Connes' embedding conjecture and tries to tell something interesting about M. In particular, he managed to associate an invariant to M, now called Brown's invariant, that carries information about rigidity properties of M. The purpose of this section is to introduce the reader to this invariant.

3.6.1 Convex Combinations of Representations into $\mathcal{R}^{\mathcal{U}}$

Let M be a separable II_1 factor verifying Connes' embedding conjecture and fix a free ultrafilter $\mathcal{U}$ on the natural numbers. The set $\mathbb{H}om(M, \mathcal{R}^{\mathcal{U}})$ of unital morphisms $M \to \mathcal{R}^{\mathcal{U}}$ modulo unitary equivalence is non-empty. We shall show that this set, that is in fact Brown's invariant, has a surprisingly rich structure.

We can equip $\mathbb{H}om(M, \mathcal{R}^{\mathcal{U}})$ with a metric in a reasonably simple way. Since M is separable, it is topologically generated by countably many elements $a_1, a_2 \ldots$, that we may assume to be contractions, that is $||a_i|| \leq 1$, for all i. So we can define a metric on $\mathbb{H}om(M, \mathcal{R}^{\mathcal{U}})$ as follows

$$d([\pi], [\rho]) = \inf_{u \in U(\mathcal{R}^{\mathcal{U}})} \left(\sum_{n=1}^{\infty} \frac{1}{2^{2n}} ||\pi(a_n) - u\rho(a_n)u^*||_2^2 \right)^{\frac{1}{2}},$$

since the series in the right hand side is convergent. A priori, d is just a pseudo-metric, but we can use Theorem 3.1 in [137] to say that approximately unitary equivalence is the same as unitary equivalence in separable subalgebras of $\mathcal{R}^{\mathcal{U}}$. This means that d is actually a metric. Moreover, while this metric may depend on the generating set $\{a_1, a_2, \ldots\}$, the induced topology does not. It is indeed the point-wise convergence topology.

$\mathbb{H}om(M, \mathcal{R}^{\mathcal{U}})$ does not carry any evident vector space structure, but Nate Brown's intuition was that one can still do convex combinations inside $\mathbb{H}om(M, \mathcal{R}^{\mathcal{U}})$ in a formal way. There is indeed an obvious (and wrong) way to proceed: given *-homomorphisms $\pi, \rho: M \to \mathcal{R}^{\mathcal{U}}$ and $0 < t < 1$, take a projection $p_t \in (\pi(M) \cup \rho(M))' \cap \mathcal{R}^{\mathcal{U}}$ such that $\tau(p_t) = t$ and define the "convex combination" $t\pi + (1 - t)\rho$ to be

$$x \mapsto \pi(x)p_t + \rho(x)p_t^{\perp}.$$

Since the projection p_t is chosen in $(\pi(M) \cup \rho(M))'$, then $t\pi + (1-t)\rho$ is certainly a new unital morphism of M in $\mathcal{R}^{\mathcal{U}}$. Unfortunately this procedure is not well defined on classes in $\mathrm{Hom}(M, \mathcal{R}^{\mathcal{U}})$ and the reason can be explained as follows: if $p \in \mathcal{R}^{\mathcal{U}}$ is a nonzero projection, then the corner $p\mathcal{R}^{\mathcal{U}}p$ is still a hyperfinite II_1-factor and so, by uniqueness, it is isomorphic to $\mathcal{R}^{\mathcal{U}}$. Thus the cut-down $p\pi$ can be seen as a new morphism $M \to \mathcal{R}^{\mathcal{U}}$. The problem is that the isomorphism $p\mathcal{R}^{\mathcal{U}}p \to \mathcal{R}^{\mathcal{U}}$ is not canonical and this reflects on the fact that convex combinations as defined above are not well-defined on classes in $\mathrm{Hom}(M, \mathcal{R}^{\mathcal{U}})$. The idea is to allow only particular isomorphisms $p\mathcal{R}^{\mathcal{U}}p \to \mathcal{R}^{\mathcal{U}}$ that are somehow fixed by conjugation by a unitary. This is done by using the so-called standard isomorphisms, that represent Nate Brown's main technical innovation.

Definition 3.6.1 Let $p \in \mathcal{R}^{\mathcal{U}}$ be a non-zero projection. A *standard isomorphism* is any map $\theta_p : p\mathcal{R}^{\mathcal{U}}p \to \mathcal{R}^{\mathcal{U}}$ constructed as follows. Lift p to a projection $(p_n) \in \ell^\infty(\mathcal{R})$ such that $\tau_{\mathcal{R}}(p_n) = \tau_{\mathcal{R}^{\mathcal{U}}}(p)$, for all $n \in \mathbb{N}$, fix isomorphisms $\theta_n : p_n\mathcal{R}p_n \to \mathcal{R}$, and define θ_p to be the isomorphism on the right hand side of the following commutative diagram

$$
\begin{array}{ccc}
\ell^\infty(p_n\mathcal{R}p_n) & \longrightarrow & p\mathcal{R}^{\mathcal{U}}p \\
\Big\downarrow {\scriptstyle \oplus \theta_n} & & \Big\downarrow {\scriptstyle \cong} \\
\ell^\infty(\mathcal{R}) & \longrightarrow & \mathcal{R}^{\mathcal{U}}
\end{array}
$$

Definition 3.6.2 Given $[\pi_1], \ldots, [\pi_n] \in \mathrm{Hom}(N, \mathcal{R}^{\mathcal{U}})$ and $t_1, \ldots, t_n \in [0, 1]$ such that $\sum t_i = 1$, we define

$$
\sum_{i=1}^{n} t_i[\pi_i] := \left[\sum_{i=1}^{n} \left(\theta_i^{-1} \circ \pi_i \right) \right],
$$

where $\theta_i : p_i\mathcal{R}^{\mathcal{U}}p_i \to \mathcal{R}^{\mathcal{U}}$ are standard isomorphisms and $p_1, \ldots, p_n \in \mathcal{R}^{\mathcal{U}}$ are orthogonal projections such that $\tau(p_i) = t_i$ for $i \in \{1, \ldots, n\}$.

We can explain in a few words why this procedure of using standard isomorphisms works. It has been originally proven by Murray and von Neumann that there is a unique unital embedding of $M_n(\mathbb{C})$ into $\mathcal{R}$ up to unitary equivalence. Since $\mathcal{R}$ contains an increasing chain of matrix algebras whose union is weakly dense, it follows that all unital endomorphisms of $\mathcal{R}$ are approximately inner. Now, if we take an automorphism Θ of $\mathcal{R}^{\mathcal{U}}$ that can be lifted (i.e. it is of the form $(\theta_n)_{n\in\mathbb{N}}$ where θ_n is an automorphism of $\ell^\infty(\mathcal{R})$), it follows that Θ is just the conjugation by some unitary, when restricted to a separable subalgebras or $\mathcal{R}^{\mathcal{U}}$. Now, Nate Brown's standard isomorphisms are exactly those isomorphisms $p\mathcal{R}^{\mathcal{U}}p \to \mathcal{R}^{\mathcal{U}}$ that are liftable and therefore it is intuitively clear that, after passing to the quotient by the relation of unitary equivalence, the choice of the standard isomorphism should not affect the result. The formalization of this rough idea leads to the following theorem.

Theorem 3.6.3 (Brown [17]) $\sum_{i=1}^{n} t_i[\pi_i]$ *is well defined, i.e. independent of the projections* p_i, *the standard isomorphisms* θ_i *and the representatives* π_i.

To prove this result we need some preliminary observations.

Lemma 3.6.4 *Let* $p, q \in \mathcal{R}$ *be projections of the same trace and* $\theta : p\mathcal{R}p \to q\mathcal{R}q$ *be a unital *-homomorphism, that is* $\theta(p) = q$. *Then there is a sequence of partial isometries* $v_n \in \mathcal{R}$ *such that:*

1. $v_n^* v_n = p$,
2. $v_n v_n^* = q$,
3. $\theta(x) = \lim_{n \to \infty} v_n x v_n^*$,

where the limit is taken in the 2-norm.

Proof Since p, q have the same trace, we can find a partial isometry w such that $w^* w = q$ and $ww^* = p$. Consider the unital endomorphism $\theta_w : p\mathcal{R}p \to p\mathcal{R}p$ defined by $\theta_w(x) = w\theta(x)w^*$. Since $\mathcal{R}$ is hyperfinite, every endomorphism is approximately inner in the 2-norm, that is, we can find unitaries $u_n \in p\mathcal{R}p$ such that $w\theta(x)w^* = \lim_{n \to \infty} u_n x u_n^*$. Defining $v_n = w^* u_n$ completes the proof. $\square$

Proposition 3.6.5 *Assume* $p, q \in \mathcal{R}^{\mathcal{U}}$ *are projections with the same trace,* $M \subseteq p\mathcal{R}^{\mathcal{U}}p$ *is a separable von Neumann subalgebra and* $\Theta : p\mathcal{R}^{\mathcal{U}}p \to q\mathcal{R}^{\mathcal{U}}q$ *is a unital *-homomorphism. Let* $(p_i), (q_i) \in \ell^{\infty}(\mathcal{R})$ *lifts of* p *and* q, *respectively, such that* $\tau_{\mathcal{R}}(p_i) = \tau_{\mathcal{R}}(q_i) = \tau_{\mathcal{R}^{\mathcal{U}}}(p)$, *for all* $i \in \mathbb{N}$. *If there are unital *-homomorphisms* $\theta_i : p_i\mathcal{R}p_i \to q_i\mathcal{R}q_i$ *such that* $(\theta_i(x_i))$ *is a lift of* $\Theta(x)$, *whenever* $(x_i) \in \Pi p_i\mathcal{R}p_i$ *is a lift of* $x \in M$, *then there are a partial isometry* $v \in \mathcal{R}^{\mathcal{U}}$ *such that:*

1. $v^* v = p$,
2. $vv^* = q$,
3. $\Theta(x) = vxv^*$, *for all* $x \in M$.

Proof We shall prove the proposition only in the case $M = W^*(X)$ is singly generated.

Let $(x_i) \in \Pi p_i\mathcal{R}p_i$ be a lift of X. By Lemma 3.6.4, there are partial isometries $v_i \in \mathcal{R}$ such that $v_i^* v_i = p_i$, $v_i v_i^* = q_i$ and $||\theta_i(x_i) - v_i x_i v_i^*||_2 < 1/i$. Observe that $(v_i) \in \ell^{\infty}(\mathcal{R})$ drops to a partial isometry $v \in \mathcal{R}^{\mathcal{U}}$ with support p and range q. To show that $\Theta(X) = vXv^*$, fix $\varepsilon > 0$ and consider the set

$$S_{\varepsilon} = \{i \in \mathbb{N} : ||\theta_i(x_i) - v_i x_i v_i^*||_2 < \varepsilon\}.$$

This set contains the cofinite set $\{i \in \mathbb{N} : i \geq \frac{1}{\varepsilon}\}$ and therefore $S_{\varepsilon} \in \mathcal{U}$. $\square$

Exercise 3.6.6 Prove Proposition 3.6.5 in the general case. (Hint: pick the v_i's to obtain inequalities of the shape $||\theta_i(Y_i) - v_i Y_i v_i^*||_2 < 1/i$ on a finite set of Y_i's that corresponds to lifts of a finite subset of a generating set of M.)

Proof of Theorem 3.6.3 Let $\sigma_i : q_i \mathcal{R}^{\mathcal{U}} q_i \to \mathcal{R}^{\mathcal{U}}$ be standard isomorphisms, built by pairwise orthogonal projections q_i of trace t_i and assume $[\rho_i] = [\pi_i]$. By Proposition 3.6.5, applied to the standard isomorphism $\sigma_i^{-1} \circ \theta_i : p_i \mathcal{R}^{\mathcal{U}} p_i \to q_i \mathcal{R}^{\mathcal{U}} q_i$, there are partial isometries $v_i \in \mathcal{R}^{\mathcal{U}}$ such that $v_i^* v_i = p_i$, $v_i v_i^* = q_i$ and

$$v_i \left(\theta_i^{-1} \circ \pi_i \right)(x) v_i^* = \left(\sigma_i^{-1} \circ \pi_i \right)(x), \qquad \text{for all } x \in M.$$

Now since $[\pi_i] = [\rho_i]$, we can find unitaries u_i such that $\rho_i = u_i \pi_i u_i^*$. The proof is then completed by the following exercise.

Exercise 3.6.7 Show that $u := \sum \sigma_i^{-1}(u_i) v_i$ is a unitary conjugating $\sum \theta_i^{-1} \circ \pi_i$ over to $\sum \sigma_i^{-1} \circ \rho_i$.

$\square$

Having a notion of convex combinations on $\mathbb{H}\mathrm{om}(N, \mathcal{R}^{\mathcal{U}})$, it is natural to ask (a) whether this set can be embedded into a vector space; and, if so, (b) what can be done with this vector space. Nate Brown himself proved in [17] that his notion of convex combinations verifies the axioms of a so-called convex-like structure. Afterwards, Capraro and Fritz showed in [22] that every convex-like structure is linearly and isometrically embeddable into a Banach space, but their proof is strongly based on Stone's representation theorem [138], which provides an abstract embedding into a vector space. Therefore, taken together, these two results provide only an abstract embedding of $\mathbb{H}\mathrm{om}(N, \mathcal{R}^{\mathcal{U}})$ into a Banach space. In the appendix, the interested reader can find a sketch of a concrete embedding, using a construction appeared in [18].

3.6.2 *Extreme Points of* $\mathbb{H}\mathrm{om}(M, \mathcal{R}^{\mathcal{U}})$ *and a Problem of Popa*

In this section we present one application of the $\mathbb{H}\mathrm{om}$ space. Given a separable II_1 factor M that embeds into $\mathcal{R}^{\mathcal{U}}$, Sorin Popa asked whether there is always another representation π such that $\pi(M)' \cap \mathcal{R}^{\mathcal{U}}$ is a factor. The following theorem by Nate Brown[17] shows that this problem is equivalent to a geometric problem on $\mathbb{H}\mathrm{om}(M, \mathcal{R}^{\mathcal{U}})$.

Theorem 3.6.8 *Let* $\pi : M \to \mathcal{R}^{\mathcal{U}}$ *be a representation. Then* $\pi(M)' \cap \mathcal{R}^{\mathcal{U}}$ *is a factor if and only if* $[\pi]$ *is an extreme point of* $\mathbb{H}\mathrm{om}(M, \mathcal{R}^{\mathcal{U}})$.

In this section we prove only the "only if" part: if $[\pi]$ is an extreme point, then $\pi(M)' \cap \mathcal{R}^{\mathcal{U}}$ is a factor.

Definition 3.6.9 We define the *cutdown* of a representation $\pi : M \to \mathcal{R}^{\mathcal{U}}$ by a projection $p \in \pi(M)' \cap \mathcal{R}^{\mathcal{U}}$ to be the map $M \to \mathcal{R}^{\mathcal{U}}$ defined by $x \to \theta_p(p\pi(x))$, where $\theta_p : p\mathcal{R}^{\mathcal{U}} p \to \mathcal{R}^{\mathcal{U}}$ is a standard isomorphism.

Lemma 3.3.3 in [17] shows that this definition is independent by the standard isomorphism, hence we can denote it by $[\pi_p]$.

Lemma 3.6.10 *Let* $\pi : M \to \mathcal{R}^{\mathcal{U}}$ *be a morphism* $p, q \in \pi(M)' \to \mathcal{R}^{\mathcal{U}}$ *be projections of the same trace. The following statement are equivalent:*

1. $[\pi_p] = [\pi_q]$,
2. *p and q are Murray-von Neumann equivalent in $\pi(M)' \cap \mathcal{R}^{\mathcal{U}}$, that is, there is a partial isometry $v \in \pi(M)' \cap \mathcal{R}^{\mathcal{U}}$ such that $v^*v = p$ and $vv^* = q$.*
3. *there exists $v \in \mathcal{R}^{\mathcal{U}}$ such that $v^*v = p$, $vv^* = q$ and $v\pi(x)v^* = q\pi(x)$, for all $x \in M$.*

Exercise 3.6.11 Prove the equivalence between (2) and (3) in Lemma 3.6.10.

Exercise 3.6.12 Given projections p, q and a partial isometry v such that $v^*v = p$ and $vv^* = q$, show that there exist lifts $(p_n), (q_n), (v_n) \in \ell^\infty(R)$ such that p_n, q_n are projections of the same trace as p, and $v_n^* v_n = p_n$, $v_n v_n^* = q_n$, for all $n \in \mathbb{N}$.

Proof of (3) $\Rightarrow$ (1) Let p_n, q_n, v_n as in Exercise 3.6.12 and fix isomorphisms $\theta_n : p_n \mathcal{R} p_n \to \mathcal{R}$ and $\gamma_n : q_n \mathcal{R} q_n \to \mathcal{R}$ and use them to define standard isomorphisms $\theta : p\mathcal{R}^{\mathcal{U}} p \to \mathcal{R}^{\mathcal{U}}$ and $\gamma : q\mathcal{R}^{\mathcal{U}} q \to \mathcal{R}^{\mathcal{U}}$ and use them to define π_p and π_q. The isomorphism on the right hand side of the following diagram[2] is liftable by construction and so Proposition 3.6.5 can be applied to it, giving unitary equivalence between π_p and π_q.

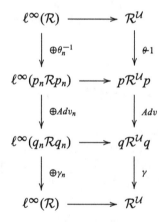

□

Exercise 3.6.13 Use a similar idea to prove the implication (1) $\Rightarrow$ (3).

We recall that a projection $p \in M$ is called *minimal* if $pMp = \mathbb{C}1$. A von Neumann algebra without minimal projections is called *diffuse*.

---

[2]The notation *Adu* in the diagram stands for the conjugation by the unitary operator u.

Exercise 3.6.14 Let M be a diffuse von Neumann algebra with the following property: every pair of projections with the same trace are Murray-von Neumann equivalent. Show that M is factor. (Hint: show that every central projection is minimal.)

Proof of the "only if" of Theorem 3.6.8 Let $[\pi]$ be an extreme point of $\mathbb{H}om$ $(M, \mathcal{R}^{\mathcal{U}})$ and $p \in \pi(M)' \cap \mathcal{R}^{\mathcal{U}}$. Since

$$[\pi] = \tau(p)[\pi_p] + \tau(p^{\perp})[\pi_{p^{\perp}}],$$

it follows that $[\pi_p] = [\pi]$, for all $p \in \pi(M)' \cap \mathcal{R}^{\mathcal{U}}$, $p \neq 0$. By Lemma 3.6.10, it follows that two projections in $\pi(M)' \cap \mathcal{R}^{\mathcal{U}}$ are Murray-von Neumann equivalent into $\pi(M)' \cap \mathcal{R}^{\mathcal{U}}$ if and only if they have the same trace. Since $\pi(M)' \cap \mathcal{R}^{\mathcal{U}}$ is diffuse, Exercise 3.6.14 completes the proof. $\qquad\square$

From Theorem 3.6.8 we obtain the following geometric reformulation of Popa's question.

Problem 3.6.15 (Geometric Reformulation of Popa's Question) Does $\mathbb{H}om$ $(M, \mathcal{R}^{\mathcal{U}})$ have extreme points?

This problem is still open. There are two obvious ways to try to attack it, leading to two related problems, whose positive solution would imply a positive solution of Popa's question.

1. Since $\mathbb{H}om(M, \mathcal{R}^{\mathcal{U}})$ is a bounded, closed and convex subset of a Banach space one cannot apply Krein-Milman's theorem and conclude existence of extreme points. Nevertheless, one can ask the question whether the Banach space into which $\mathbb{H}om(M, \mathcal{R}^{\mathcal{U}})$ embeds is actually a dual Banach space. In this case, $\mathbb{H}om(M, \mathcal{R}^{\mathcal{U}})$ would be compact in the weak*-topology and one could apply Krein-Milman's theorem.

Problem 3.6.16 Does $\mathbb{H}om(M, \mathcal{R}^{\mathcal{U}})$ embed into a dual Banach space?

Note that one way to try to attack this problem is by observing that M is the dual Banach space (every von Neumann algebra is a dual Banach space) of a unique Banach space, usually denoted by M_*. Consequently there might be some possibility to express representations $M \to \mathcal{R}^{\mathcal{U}}$ in terms of dual representations of $\mathcal{R}^{\mathcal{U}}_*$ into M_*.

We mention that Chirvasitu [95] has shown that $\mathbb{H}om(M, \mathcal{R}^{\mathcal{U}} \bar{\otimes} B(H))$ is Dedekind-complete with respect to the order induced by the cone $\mathbb{H}om_+(M, \mathcal{R}^{\mathcal{U}} \bar{\otimes} B(H))$. Since Dedekind-completeness is a necessary condition for Banach spaces of the form $C_{\mathbb{R}}(K)$, with K compact Hausdorff, to have a predual, we can consider Chirvasitu's result as a small measure of evidence that $\mathbb{H}om(M, \mathcal{R}^{\mathcal{U}} \bar{\otimes} B(H))$ is a dual Banach space.

2. The second approach is through a simple observation about geometry of Banach spaces. Recall that a Banach space B is called *strictly convex* if $b_1 \neq b_2$ and $||b_1|| = ||b_2|| = 1$ together imply that $||b_1 + b_2|| < 2$.

Exercise 3.6.17 Let B be a strictly convex Banach space and $C \subseteq B$ be a convex subset. Fix $c_0 \in C$ and assume that there is $c \in C$ such that $d(c_0, c)$ is maximized in C. Show that c is an extreme point of C.

Exercise 3.6.18 Let $M = W^*(X)$ be a singly generated II_1 factor which embeds into $\mathcal{R}^{\mathcal{U}}$. Fix $[\pi_0] \in \mathbb{H}\mathrm{om}(M, \mathcal{R}^{\mathcal{U}})$. Show that the function $\mathbb{H}\mathrm{om}(M, \mathcal{R}^{\mathcal{U}}) \ni [\pi] \to d([\pi_0], [\pi])$ attains its maximum.

With a little bit more effort one can extend Exercise 3.6.18 to every separable II_1 factor. Consequently, Exercises 3.6.17 and 3.6.18 together imply that if the convex-like structure on $\mathbb{H}\mathrm{om}(M, \mathcal{R}^{\mathcal{U}})$ were strictly convex, then Popa's question would have affirmative answer.

Problem 3.6.19 Does $\mathbb{H}\mathrm{om}(M, \mathcal{R}^{\mathcal{U}})$ embed into a strictly convex Banach space?

Observe that Theorem 3.6.21 suggests that $\mathbb{H}\mathrm{om}(M, \mathcal{R}^{\mathcal{U}})$ itself should not be strictly convex.

The study of the extreme points of $\mathbb{H}\mathrm{om}(M, \mathcal{R}^{\mathcal{U}})$ is not only interesting in light of Popa's question, but also because it provides a method to distinguish II_1 factors. For instance, Brown proved in [17], Corollary 5.4, that rigidity of an $\mathcal{R}^{\mathcal{U}}$-embeddable II_1 factor M with property (T) reflects on the rigidity of the set of the extreme points of its $\mathbb{H}\mathrm{om}(M, \mathcal{R}^{\mathcal{U}})$, that turns out to be discrete. Property (T) for von Neumann algebras is a form of rigidity introduced by Connes and Jones in [44] and inspired to Kazhdan's property (T) for groups [98]. A simple way to define property (T) for von Neumann algebras is through the following definition.

Definition 3.6.20 A II_1 factor M with trace τ has *property (T)* if for all $\varepsilon > 0$, there exist $\delta > 0$ and a finite subset F of M such that for all τ-preserving ucp maps $\phi : M \to M$, one has

$$\sup_{x \in F} ||\phi(x) - x||_2 \leq \delta \Rightarrow \sup_{Ball(M)} ||\phi(x) - x||_2 \leq \varepsilon.$$

The interpretation of property (T) as a form of rigidity should be clear: if a trace-preserving ucp map is closed to the identity on a finite set, then it is actually close to the identity on the whole von Neumann algebra.

Classical examples of factors with property (T) are the ones associated to $SL(n, \mathbb{Z})$, with $n \geq 3$. The following result was proved by Nate Brown in [17], Corollary 5.4.

Theorem 3.6.21 *If M has property (T), then the set of extreme points of* $\mathbb{H}\mathrm{om}(M, \mathcal{R}^{\mathcal{U}})$ *is discrete.*

Proof Popa proved in [127], Sect. 4.5, that for every $\varepsilon > 0$, there is a $\delta > 0$ such that if $[\pi], [\rho] \in \mathbb{H}\mathrm{om}(M, \mathcal{R}^{\mathcal{U}})$ are at distance $\leq \delta$, then there are projections $p \in \pi(M)' \cap \mathcal{R}^{\mathcal{U}}$ and $q \in \rho(M) \cap \mathcal{R}^{\mathcal{U}}$ and a partial isometry v such that $v^*v = p$, $vv^* = q$, $\tau(p) > 1 - \varepsilon$, and $v\pi(x)v^* = q\rho(x)$, for all $x \in M$. This implies that

$p\pi$ and $q\rho$ are approximately unitarily equivalent and consequently, by countable saturation, $[\pi_p] = [\rho_q]$.

Now fix $\varepsilon > 0$ assume that $[\pi]$ and $[\rho]$ are δ-close extreme points and take projections $p \in \pi(M)' \cap \mathcal{R}^\mathcal{U}$ and $q \in \rho(M)' \cap \mathcal{R}^\mathcal{U}$ such that $[\pi_p] = [\rho_q]$. Since $[\pi]$ and $[\rho]$ are extreme points, we can apply Theorem 3.6.8 and conclude that $[\pi] = [\pi_p]$ and $[\rho] = [\rho_q]$, that is, $[\pi] = [\rho]$. $\qquad\square$

Remark 3.6.22 Observe that Theorem 3.6.21 tells that the set of extreme points is discrete but, as far as we know, it might be empty. Indeed the problem of proving that extreme points actually exist for $\mathcal{R}^\mathcal{U}$-embeddable II$_1$ factors is open even for factors with property (T).

We conclude mentioning that also examples of factors with a continuous non-empty set of extreme points are also known in (see [17] Corollaries 6.10 and 6.11).

Conclusions

The problem of whether every countable discrete group is sofic or hyperlinear is currently of paramount importance. Along this monograph, we have shown that answering these problems would automatically settle a number of conjectures from different fields of pure and applied mathematics.

In the sofic case this problem seems to boil down to understanding which relations (or, more generally, existential formulas in the language of invariant length groups; see Sect. 2.6) are *approximately satisfiable* in the permutation groups endowed with the Hamming distance. This connection has been made explicit in [7, 71]. Similar arguments holds for hyperlinear groups.

Suppose that $w(\bar{x})$ is a word, where $\bar{x} = (x_1, \ldots, x_n)$, and G is an invariant length group. A tuple $\bar{g} = (g_1, \ldots, g_n)$ in G is a *solution* of the equation $\ell(w(\bar{x})) = 0$ if $\ell_G(\bar{w}(\bar{g})) = 0$ or, equivalently, $w(\bar{g})$ equals the identity of G. A δ-*approximate solution* of $\ell(w(\bar{x})) = 0$ is a tuple $(\bar{g})$ such that $\ell^G(w(\bar{g})) < \delta$.

Suppose now that C is a class of invariant length groups. The formula $\ell(w(\bar{x}))$ is *stable* with respect to C if, roughly speaking, every approximated solution of such an equation is close to an exact solution. More precisely for any given $\varepsilon > 0$ there is $\delta > 0$ such that whenever G is an element of the class C and $\bar{g}$ is δ-approximate solution of the equation $\ell(w(\bar{x})) = 0$ in G, there is an exact solution $\bar{h}$ in G such that $\max_i \ell_G(g_i h_i^{-1}) < \varepsilon$.

More generally if $\varphi(\bar{x})$ is an arbitrary formula in the language of invariant length groups, then one can define as above the notion of solution and δ-approximate solution of the equation $\varphi(\bar{x}) = 0$. The formula $\varphi(\bar{x})$ is *stable* with respect to a class C if for any $\varepsilon > 0$ there is $\delta > 0$ such that whenever G is an element of C and $\bar{g}$ is a tuple in G such that $\varphi(\bar{g}) < \delta$ there is a tuple $\bar{h}$ in G such that $\varphi(\bar{h}) = 0$ and $\max_i \ell_G(g_i h_i^{-1}) < \varepsilon$. Stability is a key notion in model theory for metric structures, being tightly connected to the concept of definability [27, 57]. (This use of the word "stability" should not be confused with the classical model-theoretic stability theory as in [135].) In the field of operator algebras, stability and the related notion of (weak) semiprojectivity are of fundamental importance; see [107].

© Springer International Publishing Switzerland 2015

V. Capraro, M. Lupini, *Introduction to Sofic and Hyperlinear Groups and Connes' Embedding Conjecture*, Lecture Notes in Mathematics 2136, DOI 10.1007/978-3-319-19333-5

Recall from Sect. 2.9 that a countable discrete group has the *C-approximation property* if it admits a *length-preserving* embedding into a length ultraproduct of elements of C, i.e. an embedding such that the image of every element other than the identity has length 1. For a discrete group, soficity is equivalent to the C-approximation property, where C is the class of permutation groups endowed with the Hamming distance. Similarly a discrete group is LEF (respectively LEA) if and only if it has the C-approximation property, where C is the class of finite (resp. amenable) groups endowed with the trivial length function.

Stability of a formula with respect to a class C of invariant length groups allows one to perturb a C-approximation to obtain a C_{disc}-approximation, where C_{disc} is the class of groups from C *endowed with the trivial length function*. The following proposition is a consequence of this observation.

Proposition *Suppose that*

$$\Gamma = \langle g_1, \ldots, g_n : w_i(\bar{g}) = 1 \text{ for } i \leq k \rangle$$

is a finitely presented group. If Γ has the C-approximation property, and the formula

$$\max_{i \leq k} \ell\left(w_i(\bar{x})\right) + \left(1 - \min_{j \leq n} \ell\left(x_j\right)\right)$$

is stable with respect to the class C, then Γ has the C_{disc}-approximation property.

In the particular case when C is the class of permutation groups endowed with the Hamming length function one obtains the following consequence; see also [71, Proposition 3] and [7, Theorem 7.3]. (Recall that a finitely presented LEF group is residually finite.)

Corollary (Glebsky-Rivera, Arzhantseva-Paunsecu) *Suppose that*

$$\Gamma = \langle g_1, \ldots, g_n : w_i(\bar{g}) = 1 \text{ for } i \leq n \rangle$$

is a finitely-presented group that is not residually finite. If the formula

$$\max_{i \leq n} \ell\left(w_i(\bar{x})\right) + 1 - \min_{j \leq n} \ell\left(x_j\right)$$

is stable with respect to the class of permutation groups endowed with the Hamming length function, then Γ is not sofic.

This provides a possible line of attack to the soficity problem of some groups, first suggested in [7, 71]. There are in fact many finitely-presented groups that are known to be not residually finite. Among these Higman's group which we have considered in Sect. 2.9. Recall that this is the group H with presentation

$$\langle h_1, h_2, h_3, h_4 : h_{i+1}h_ih_{i+1}^{-1}h_i^{-2} = 1 \text{ for } i \leq 4 \rangle$$

where the sum $i + 1$ is evaluated modulo 4. Thus the corollary above provides the following sufficient conditions for H being not sofic: the formula

$$\max_{i \leq k} \ell \left(x_{i+1} x_i x_{i+1}^{-1} x_i^{-2} \right) + 1 - \min \{ \ell (x_1) , \ell (x_2) , \ell (x_3) , \ell (x_4) \}$$

is stable with respect to the class of permutation groups endowed with the Hamming length function. This means that for every $\varepsilon > 0$ there is $\delta > 0$ such that whenever $n \in \mathbb{N}$ and $\sigma_1, \sigma_2, \sigma_3, \sigma_4 \in S_n$ are such that $\ell_{S_n} (\sigma_i) > 1 - \delta$ and

$$\ell_{S_n} \left(\sigma_{i+1} \sigma_i \sigma_{i+1}^{-1} \sigma_i^{-2} \right) < \delta$$

for $i \leq 4$, then there are $\tau_1, \tau_2, \tau_3, \tau_4 \in S_n$ such that $\tau_{i+1} \tau_i \tau_{i+1}^{-1} = \tau_i^2$ and $\ell_{S_n} \left(\tau_i \sigma_i^{-1} \right) < \varepsilon$ for $i \leq 4$.

Another candidate for this line of approach is Thompson's group F. This is a very famous groups with several equivalent descriptions. For our purposes it can be regarded as the finitely presented groups with presentation

$$\langle a, b : \left[ab^{-1}, a^{-1}ba \right] = \left[ab^{-1}, a^{-2}ba^2 \right] = 1 \rangle.$$

While it is known that F is not elementarily amenable, it is a famous open problem whether F is amenable. If fact it even seems to be unknown whether F is sofic. In view of the corollary above one can conclude that a sufficient condition to refute soficity of F is the stability of the formula

$$\max \{ \ell \left(\left[xy^{-1}, x^{-1}yx \right] \right) , \ell \left(\left[xy^{-1}, x^{-2}yx^2 \right] \right) \} + 1 - \min \{ \ell (x) , \ell (y) \}$$

with respect to the class of permutation groups endowed with the Hamming length function. Concretely this means that for every $\varepsilon > 0$ there is $\delta > 0$ such that whenever $n \in \mathbb{N}$ and $\sigma, \rho \in S_n$ are such that $\ell_{S_n} (\sigma) > 1 - \delta$, $\ell_{S_n} (\rho) > 1 - \delta$, $\ell_{S_n} \left(\left[\sigma \rho^{-1}, \sigma^{-1} \rho \sigma \right] \right) < \delta$, and $\ell_{S_n} \left(\left[\sigma \rho^{-1}, \sigma^{-2} \rho \sigma^2 \right] \right) < \delta$, there are $\tau, \lambda \in S_n$ such that $\tau \lambda^{-1}$ commutes with $\tau^{-1} \lambda \tau$ and $\tau^{-2} \lambda \tau^2$, $\ell_{S_n} \left(\sigma \tau^{-1} \right) < \varepsilon$, and $\ell_{S_n} \left(\rho \lambda^{-1} \right) < \varepsilon$.

These reformulations show the importance of determining stability of formulas in permutation groups. This problem seems to be currently not well understood. Among the few papers dedicated to this subject we can mention [7, 71, 114]. In particular it is shown in [7] that, remarkably, the commutator relation $\ell \left(xyx^{-1}y^{-1} \right) = 0$ is stable in permutation groups. This seems to be the first natural step towards determining whether the nonsoficity criterion above applies to Higman's and Thompson's group.

Appendix A
Tensor Product of C*-Algebras

A *normed *-algebra* is an algebra A (over $\mathbb{C}$) equipped with:

1. an involution $*$ such that

   - $(x + y)^* = x^* + y^*$,
   - $(xy)^* = x^* y^*$,
   - $(\lambda x)^* = \bar{\lambda} x^*$.

2. a norm $|| \cdot ||$ such that $||xy|| \leq ||x|| ||y||$.

   A Banach *-algebra is a normed *-algebra that is complete.

Definition A.1 A *C*-algebra* is a Banach *-algebra verifying the following additional property, called *C*-identity*:

$$||x^*x|| = ||x||^2, \qquad \forall a \in A.$$

The C*-identity is a relation between algebraic and topological properties. It indeed implies that sometimes algebraic properties imply topological properties. A classical example of this interplay is the following fact.

Fact A *-homomorphism from a normed *-algebra to a C*-algebra is always a contraction.

Example of C*-algebras certainly include the (commutative) algebra of complex valued functions on a compact space equipped with the sup norm and $B(H)$ itself. But, exactly as in case of von Neumann algebras, one can start from a group and construct a C*-algebra in a natural way.

Example A.2 Let G be a locally compact group. Fix a left-Haar measure μ and construct the convolution *-algebra $L^1(G)$ as follows: the elements of $L^1(G)$ are μ-integrable complex-valued functions on G. The convolution is defined by

© Springer International Publishing Switzerland 2015

V. Capraro, M. Lupini, *Introduction to Sofic and Hyperlinear Groups and Connes' Embedding Conjecture*, Lecture Notes in Mathematics 2136, DOI 10.1007/978-3-319-19333-5

$(f * g)(x) = \int_G f(y)g(y^{-1}x)d\mu(y)$ and the involution is defined by $f^*(x) = \overline{f(x^{-1})}\Delta(x^{-1})$, where Δ is the modular function, i.e. the (unique) function $\Delta : G \to [0, \infty)$ such that for all Borel subsets A of G one has $\mu(Ax^{-1}) = \Delta(x)\mu(A)$. The *full C*-algebra* of G, denoted by $C^*(G)$ is the enveloping C*-algebra of $L^1(G)$, i.e. the completion of $L^1(G)$ with respect to the norm $||f|| = sup_\pi ||\pi(f)||$, where π runs over all non-degenerate *-representations of $L^1(G)$ in a Hilbert space.[1]

Exercise A.3   Show that in fact $|| \cdot ||$ is a norm on $L^1(G)$.

In particular, we can make this construction for the free group on countably many generators, usually denoted by $\mathbb{F}_\infty$. Observe that this group is countable and so its Haar measure is the counting measure which is bi-invariant. Consequently, the modular function is constantly equal to 1.

Exercise A.4   Let $\delta_g : G \to \mathbb{R}$ be the characteristic function of the point g. Show that $g \to \delta_g$ is an embedding $\mathbb{F}_\infty \hookrightarrow U(C^*(\mathbb{F}_\infty))$.

Exercise A.5   Prove that the unitaries in Exercise A.4 form a norm total sequence in $C^*(\mathbb{F}_\infty)$.

Given two C*-algebras A_1 and A_2, their algebraic tensor product is a *-algebra in a natural way, by setting

$$(x_1 \otimes x_2)(y_1 \otimes y_2) = x_1x_2 \otimes y_1y_2,$$

$$(x_1 \otimes x_2)^* = x_1^* \otimes x_2^*.$$

Nevertheless it is not clear how to define a norm to obtain a C*-algebra.

Definition A.6   Let A_1, A_2 be two C*-algebras and $A_1 \odot A_2$ their algebraic tensor product. A norm $|| \cdot ||_\beta$ on $A_1 \otimes A_2$ is called C*-*norm* if the following properties are satisfied:

1. $||xy||_\beta \leq ||x||_\beta ||y||_\beta$, for all $x, y \in A_1 \odot A_2$;
2. $||x^*x||_\beta = ||x||_\beta^2$, for all $x \in A_1 \odot A_2$.

If $|| \cdot ||_\beta$ is a C*-norm on $A_1 \odot A_2$, then $A_1 \otimes_\beta A_2$ denotes the completion of $A_1 \odot A_2$ with respect to $|| \cdot ||_\beta$. It is a C*-algebra.

Exercise A.7   Prove that every C*-norm β is multiplicative on elementary tensors, i.e. $||x_1 \otimes x_2||_\beta = ||x_1||_{A_1}||x_2||_{A_2}$.

The interesting thing is that one can define at least two different C*-norms, a minimal one and a maximal one, and they are indeed different in general. To define these norms, let us first recall that a representation of a *-algebra A is a *preserving

---

[1] A representation $\pi : L^1(G) \to B(H)$ is said to be non-degenerate if the set $\{\pi(f)\xi : f \in L^1(G), \xi \in H\}$ is dense in H.

algebra-morphism from A to some $B(H)$. We denote Rep(A) the set of representation of A.

Definition A.8

$$||x||_{max} = \sup\{||\pi(x)|| : \pi \in Rep(A_1 \odot A_2)\} \qquad (A.1)$$

Exercise A.9 Prove that $||\cdot||_{max}$ is indeed a C*-norm. (Hint: to prove that $||x||_{max} < \infty$, for all x, take inspiration from Lemma 11.3.3(iii) in [92].)

The norm $||\cdot||_{max}$ is named *maximal norm*, or projective norm, or Turumaru's norm, being first introduced in [146]. The completion of $A_1 \odot A_2$ with respect to it is denoted by $A_1 \otimes_{max} A_2$.

Let A be a C*-algebra and $S \subseteq A$. Denote by $C^*(S)$ the C*-subalgebra of A generated by S. The maximal norm has the following universal property (see [140], IV.4.7).

Proposition A.10 *Given C*-algebras A_1, A_2, B. Assume $\pi_i : A_i \to B$ are *-homomorphisms with commuting ranges, that is for all $x \in \pi_1(A_1)$ and $y \in \pi_2(A_2)$, one has $xy = yx$. Then there exists a unique *-homomorphism $\pi : A_1 \otimes_{max} A_2 \to B$ such that*

1. $\pi(x_1 \otimes x_2) = \pi_1(x_1)\pi_2(x_2)$
2. $\pi(A_1 \otimes_{max} A_2) = C^*(\pi_1(A_1), \pi_2(A_2))$

Exercise A.11 Prove Proposition A.10.

We now turn to the definition of the minimal C*-norm. The idea behind its definition is very simple. Instead of considering all representations of the algebraic tensor product, one takes only representations which split into the tensor product of representations of the factors.

Definition A.12

$$||x||_{min} = \sup\{||(\pi_1 \otimes \pi_2)(x)|| : \pi_i \in Rep(A_i)\} \qquad (A.2)$$

Exercise A.13 Prove that $||\cdot||_{min}$ is a C*-norm on $A_1 \odot A_2$.

This norm is named *minimal norm*, or injective norm, or Guichardet's norm, being first introduced in [81]. The completion of $A_1 \odot A_2$ with respect to it is denoted by $A_1 \otimes_{min} A_2$.

Remark A.14 Clearly $||\cdot||_{min} \leq ||\cdot||_{max}$, since representations of the form $\pi_1 \otimes \pi_2$ are particular *-representation of the algebraic tensor product $A_1 \odot A_2$. These norms are different, in general, as Takesaki showed in [139]. Notation $||\cdot||_{max}$ reflects the fact that there are no C*-norms greater than the maximal norm, that follows straightforwardly from the GNS construction. Notation $||\cdot||_{min}$ has the same justification, but it is much harder to prove:

Theorem A.15 (Takesaki, [139]) $||\cdot||_{min}$ *is the smallest C*-norm on $A_1 \odot A_2$.*

Appendix B
Ultrafilters and Ultralimits

Vladimir G. Pestov

B.1 Let us recall that the symbol ℓ^∞ denotes the linear space consisting of all bounded sequences of real numbers. This linear space is sometimes viewed as an algebra, meaning that it supports a natural multiplication (the product of two bounded sequences is bounded). Besides, ℓ^∞ is equipped with the supremum norm,

$$\|x\| = \sup_{n=1}^{\infty}|x_n|$$

and the corresponding metric.

B.2 The usual notion of the limit of a convergent sequence of real numbers can be interpreted as a mapping from a subset of ℓ^∞ consisting of all convergent sequences (this subset is usually denoted c, it is a closed normed subspace of ℓ^∞) to $\mathbb{R}$. This map,

$$c \ni x = (x_n)_{n=1}^{\infty} \mapsto \lim_{n\to\infty} x_n \in \mathbb{R},$$

has the following well-known properties (refer to first-year Calculus for their proofs):

1. $\lim_{n\to\infty}(x_n + y_n) = \lim_{n\to\infty} x_n + \lim_{n\to\infty} y_n$.

V.G. Pestov (✉)
Departamento de Matemática, Universidade Federal de Santa Catarina, Campus Universitário Trindade, CEP 88.040-900 Florianópolis-SC, Brasil (Pesquisador Visitante Especial do CAPES, processo 085/2012)

Department of Mathematics and Statistics, University of Ottawa, 585 King Edward Avenue, Ottawa, ON, Canada K1N6N5
e-mail: vpest283@uottawa.ca

© Springer International Publishing Switzerland 2015
V. Capraro, M. Lupini, *Introduction to Sofic and Hyperlinear Groups and Connes' Embedding Conjecture*, Lecture Notes in Mathematics 2136,
DOI 10.1007/978-3-319-19333-5

2. $\lim_{n\to\infty}(x_n y_n) = \lim_{n\to\infty} x_n \cdot \lim_{n\to\infty} y_n.$
3. $\lim_{n\to\infty}(\lambda x_n) = \lambda \lim_{n\to\infty} x_n.$
4. If $x_n \leq y_n$, then $\lim_{n\to\infty} x_n \leq \lim_{n\to\infty} y_n.$
5. The limit of a constant sequence of ones is 1:

$$\lim_{n\to\infty} \bar{1} = 1,$$

where $\bar{1} = (1, 1, 1, \ldots).$

B.3   In a sense, the main shortcoming of the notion of a limit is that it is not defined for *every* bounded sequence. For instance, the following sequence has no limit in the usual sense:

$$0, 1, 0, 1, 0, 1, 0, \ldots, 0, 1, 0, \ldots \qquad (\text{B.1})$$

As we all remember from the student years, there are very subtle cases where proving or disproving the convergence of a particular sequence can be tricky!

Can one avoid such difficulties and devise the notion of a "limit" which would have all the same properties as above, and yet with regard to which *every sequence* will have a limit, even the one in formula (B.1)?

B.4   The response to the question in (B.3) turns out to be very simple and disappointing. It is enough, for example, to define a "limit" of a sequence (x_n) by selecting the first term:

$$\phi(x) = x_1.$$

This clearly satisfies all the properties listed in (B.2). But this is not what we want: such a "limit" does not reflect the asymptotic behaviour of a sequence at the infinity. And of course this unsatisfying response is due to a poorly-formulated question. We have, therefore, to reformulate the question as follows.

B.5   Does there exist a notion of a "limit" of a sequence which

- is defined for all bounded sequences,
- has all the same properties listed in (B.2), and
- coincides with the classical limit in the case of convergent sequences?

B.6   Answering this question is the main goal of the present Appendix. Meanwhile, the sequence in Eq. (B.1) will serve as a guiding star, or rather as a guinea pig. Indeed, assuming such a wonderful specimen of a limit does exist, it will in particular assign a limit to this concrete sequence, so perhaps it would be a good idea, to begin by trying to guess what it will be? Say, will $1/2$ be a reasonable suggestion?

Let us analyse this question in the context of all *binary* sequences, that is, sequences taking values 0 and 1 only. The family of all binary sequences is $\{0, 1\}^{\mathbb{N}_+}$, the Cantor space (though the metric and topology on this space will not play any role for the moment). Every binary sequence is just a map

$$x: \mathbb{N}_+ \to \{0, 1\},$$

and so can be identified with the characteristic function of a suitable subset $A \subseteq \mathbb{N}_+$ of natural numbers, namely the set of all n where x_n takes value one:

$$\chi_A(n) = \begin{cases} 1, & \text{if } n \in A, \\ 0, & \text{otherwise.} \end{cases}$$

For instance, the sequence in Eq. (B.1) is the characteristic (indicator) function of the set of all even natural numbers.

B.7   Suppose the desired limit exists. Let us denote it, for the time being, simply by lim. What are the properties of this limit on the set of binary sequences?

First of all, it must send the constant sequence of ones (the indicator function of $\mathbb{N}_+$) to one, this was one of the rules required:

B.7.1 $\lim \chi_{\mathbb{N}_+} = \lim(1, 1, 1, 1, \ldots) = 1$.

This implies that the limit of the zero sequence (the indicator function of the empty set) is zero: indeed, the sum of the two sequences is the sequence of ones, and the additive property of the limit implies

$$1 + \lim \chi_\emptyset = \lim \chi_{\mathbb{N}_+} + \lim \chi_\emptyset = \lim \chi_{\mathbb{N}_+} = 1.$$

We summarize:

B.7.2 $\lim \chi_\emptyset = \lim(0, 0, 0, 0, \ldots) = 0$.

The next observation is that the only possible values of our limit on binary sequences are 0 or 1. Indeed, for a binary sequence (ε_n), where $\varepsilon_n \in \{0, 1\}$, one has $\varepsilon_n^2 = \varepsilon_n$, and so

$$\lim(\varepsilon_n) = \lim(\varepsilon_n^2) = (\lim \varepsilon_n)^2,$$

but there are only two real numbers satisfying the property $\varepsilon^2 = \varepsilon$.

B.7.3   For every subset $A \subseteq \mathbb{N}_+$, one has $\lim \chi_A \in \{0, 1\}$.

This excludes a "natural" possibility to assign the limit $1/2$ to the sequence in Eq. (B.1).

Now let χ_A be any binary sequence. We can "flip" the sequence and replace all ones with zeros, and vice versa. This corresponds to the indicator function of the complement $A^c = \mathbb{N}_+ \setminus A$. What about the limit of this sequence? It turns out the limit flips as well.

B.7.4 $\lim(\chi_{A^c}) = 1 - \lim \chi_A$.
   Indeed, $\chi_A + \chi_{A^c} = \chi_{\mathbb{N}_+}$ is the identical sequence of ones whose limit is one, and now one has

$$\lim \chi_{A^c} = \lim \chi_{\mathbb{N}_+} - \lim \chi_A = 1 - \lim \chi_A.$$

Since the limit of *every* binary sequence must exist by assumption, this property implies the next one:

B.7.5   Let A be any subset of the positive natural numbers. Then either $\lim \chi_A = 1$, or $\lim \chi_{A^c} = 1$, but not both at the same time.
   Let us address the following situation. Suppose we have a sequence χ_A whose limit we know to be one. Now we take a subset B of the natural numbers which contains A:

$$B \supseteq A,$$

and consider the sequence χ_B. Thus, some zeros in χ_A have been (possibly) replaced with ones, and all the ones in χ_A keep their values. What about the limit of the new sequence, χ_B?

B.7.6   If $A \subseteq B$ and $\lim \chi_A = 1$, then $\lim \chi_B = 1$.
   Indeed, $\chi_A \leq \chi_B$, and according to the property (5) in (B.2), we conclude:

$$\lim \chi_A \leq \lim \chi_B.$$

However, $\lim \chi_A = 1$, and $\lim \chi_B$ cannot be strictly greater than one. We conclude.
   The next situation to consider is as follows. Suppose A and B are two subsets of the natural numbers. Their intersection corresponds to the product of the indicator functions, as is well known and easy to see:

$$\chi_A \cdot \chi_B = \chi_{A \cap B}.$$

It follows that

$$\lim \chi_A \cdot \lim \chi_B = \lim \chi_{A \cap B}.$$

In particular:

B.7.7   If $\lim \chi_A = \lim \chi_B = 1$, then $\lim \chi_{A \cap B} = 1$.

B.8   Now it is time to put together some of the properties proved. Notice that instead of binary functions, one can just as well talk of subsets of the set $\mathbb{N}_+$. To some subsets A there is associated the value one (when $\lim \chi_A = 1$), to others, the value zero (if $\lim \chi_A = 0$). All subsets of $\mathbb{N}_+$ are therefore grouped in two classes. Let us denote the class of all subsets A to which we associate the limit one by $\mathcal{U}$:

$$\mathcal{U} = \{A \subseteq \mathbb{N}_+ : \lim \chi_A = 1\}.$$

The properties established above immediately translate into the following.

1. $\emptyset \notin \mathcal{U}$.
2. For every subset $A \subseteq \mathbb{N}_+$, either $A \in \mathcal{U}$, or $A^c \in \mathcal{U}$.
3. If $A, B \in \mathcal{U}$, then $A \cap B \in \mathcal{U}$.

B.8.1   *Ultrafilters.* A collection of subsets of natural numbers (or, in fact, of any fixed non-empty set) satisfying the above three axioms is called an *ultrafilter* (on this set).

There is no need to include more axioms in the definition of an ultrafilter.

B.8.2

Exercise Convert the rest of the properties that we have established in (B.7.1)–(B.7.7) into a set-theoretic form and deduce them from the three axioms above.

B.8.3

Example The property (B.7.6) becomes: if $A \subseteq B$ and $A \in \mathcal{U}$, then $B \in \mathcal{U}$.

◁ This follows from the axioms of an ultrafilter: assuming the contrary, one must have $\mathbb{N} \setminus B \in \mathcal{U}$ (axiom 2), meaning $\emptyset = A \cap (X \setminus B) \in \mathcal{U}$, a contradiction.   ▷

B.9   Thus, each time we have the concept of a limit with the properties that we have specified in (B.2), we get an ultrafilter on the set of natural numbers.

B.9.1   *Principal Ultrafilters.* In particular, to the disappointing example of a "limit" described in (B.4), there corresponds the following ultrafilter:

$$\mathcal{U} = \{A \subseteq \mathbb{N} : 1 \in A\}.$$

In other words, a binary sequence has limit one if and only if the first term of this sequence is one. This happens exactly when $A \ni 1$, where our sequence is χ_A.

More generally, one can repeat the construction given any chosen element $n \in \mathbb{N}_+$. The resulting ultrafilter is denoted (n),

$$(n) = \{A \subseteq \mathbb{N}_+ : n \in A\},$$

and called a *trivial* (or: *principal*) ultrafilter.

B.9.2   *Non-principal Ultrafilters.* This is, in a sense, a non-interesting situation. What is an interesting one? An ultrafilter $\mathcal{U}$ is called *free,* or *non-principal,* if its elements have no points in common:

$$\bigcap \mathcal{U} = \emptyset.$$

B.9.3   *Every Ultrafilter Is Either Trivial or Free.*
    ◁ Suppose $\mathcal{U}$ is an ultrafilter which is not trivial. This means: for every $n \in \mathbb{N}_+$, $\mathcal{U} \neq (n)$. This can mean two things: either there is $A \ni n$ which is not in the ultrafilter (in which case A^c is), or else there is $A \in \mathcal{U}$ so that $A \not\ni n$. In both cases, we can find an element A of $\mathcal{U}$ not containing n. Since this holds for all n, it is clear that the intersection of all members of $\mathcal{U}$ is an empty set, and the ultrafilter is non-principal.   ▷

B.10   How to establish the existence of free ultrafilters? We need the following notion. A system $\mathcal{C}$ of subsets of a certain non-empty set is called a *centred system* if the intersection of every finite collection of elements of $\mathcal{C}$ is non-empty:

$$\forall n, \ \ \forall A_1, A_2, \ldots, A_n \in \mathcal{C}, \ \ A_1 \cap A_2 \cap \ldots \cap A_n \neq \emptyset.$$

B.10.1   *Every Ultrafilter Is a Centred System.*
    This follows immediately from one of the axioms of an ultrafilter.

B.10.2   *Every Ultrafilter Is a Maximal Centred System. In Other Words, if $\mathcal{U}$ Is an Ultrafilter and $A \notin \mathcal{U}$, Then the System $\mathcal{U} \cup \{A\}$ Is Not Centred.*
    ◁ Indeed, since $A \notin \mathcal{U}$, we must have $A^c \in \mathcal{U}$, and since A and A^c are both members of the system $\mathcal{U} \cup \{A\}$ and their intersection is empty, the system is not centred.   ▷

B.10.3   *Ultrafilters Are Exactly Maximal Centred Systems.*
    More precisely, let $\mathscr{C}$ be a collection of subsets of a non-empty set (in our case, we are interested in the set of natural numbers, $\mathbb{N}_+$). Then $\mathscr{C}$ is an ultrafilter if and only if $\mathscr{C}$ is a maximal centred system.
    We have already established $\Rightarrow$ in the previous paragraph, now let us prove the other implication.
    ◁ Suppose $\mathscr{C}$ is a maximal centred system. "Maximal" means that if we add to $\mathscr{C}$ any subset A of $\mathbb{N}_+$ which is not in $\mathscr{C}$, then the resulting system of sets

$$\mathscr{C} \cup \{A\}$$

is no longer centred. Let us verify all three properties of an ultrafilter.

(1) Clearly, $\emptyset \notin \mathscr{C}$, because otherwise $\mathscr{C}$ would not be a centred system to begin with: for example,

$$\emptyset \cap \emptyset = \emptyset.$$

(2) Let $A, B \in \mathscr{C}$. We need to prove $A \cap B \in \mathscr{C}$. Consider the system

$$\mathscr{C} \cup \{A \cap B\}.$$

This system is centred: indeed, if $A_1, \ldots, A_n$ are elements of this system, then, if $A \cap B$ is not among them, their intersection is clearly non-empty (as $\mathscr{C}$ is centred). If we add $A \cap B$, then

$$A_1 \cap \ldots \cap A_n \cap (A \cap B) = A_1 \cap \ldots \cap A_n \cap A \cap B \neq \emptyset,$$

because $\mathscr{C}$ is centred and all sets $A_1, \ldots, A_n, A, B$ belong to $\mathscr{C}$.

We conclude: the system $\mathscr{C} \cup \{A \cap B\}$ cannot be strictly larger than $\mathscr{C}$, because of maximality of the latter. But this means that

$$A \cap B \in \mathscr{C}.$$

(3) Let $A \subseteq \mathbb{N}_+$ be any subset. We want to show that either A or its complement $A^c = \mathbb{N}_+ \setminus A$ belong to $\mathscr{C}$. Suppose neither holds, towards a contradiction. The assumption that $A \notin \mathscr{C}$ implies that the system $\mathscr{C} \cap \{A\}$ is not centred, and so there exist $A_1, A_2, \ldots, A_m \in \mathscr{C}$ with

$$A_1 \cap A_2 \cap \ldots \cap A_m \cap A = \emptyset.$$

Note that this observation can be rewritten as

$$A_1 \cap A_2 \cap \ldots \cap A_m \subseteq A^c. \tag{B.2}$$

Likewise, the assumption $A^c \notin \mathscr{C}$ implies the existence of elements $B_1, B_2, \ldots, B_k \in \mathscr{C}$ satisfying

$$B_1 \cap B_2 \cap \ldots \cap B_k \cap A^c = \emptyset,$$

or, in an equivalent form,

$$B_1 \cap B_2 \cap \ldots \cap B_k \subseteq A. \tag{B.3}$$

Together, the equations (B.2) and (B.3) imply

$$A_1 \cap A_2 \cap \ldots \cap A_m \cap B_1 \cap B_2 \cap \ldots \cap B_k \subseteq A \cap A^c = \emptyset,$$

which contradicts the fact that $\mathscr{C}$ is a centred system. We are done.   ▷

B.11   Existence of Free Ultrafilters.

B.11.1   *Zorn's Lemma. Let $(\mathfrak{X}, \leq)$ be a partially ordered set. Suppose that X has the following property: every totally ordered subset $\mathfrak{C}$ of $\mathfrak{X}$ has an upper bound. (One says that $\mathfrak{X}$ is* inductive.*) Then $\mathfrak{X}$ has a maximal element.*

An element x is *maximal* if there is no $y \in \mathfrak{X}$ which is strictly larger than x. It does not mean that x is necessarily the largest element of $\mathfrak{X}$.

The statement of Zorn's lemma is an equivalent form of the Axiom of Choice, which is a part of the standard system of axioms of set theory ZFC.

B.11.2

Exercise Let $\mathfrak{C}$ be a family of centred systems on a set $\mathbb{N}_+$ which is totally ordered by inclusion. Prove that the union of this family, $\cup\mathfrak{C}$, is again a centred system.

B.11.3   *Every Centred System Is Contained in a Maximal Centred System.*

◁ Let $\mathscr{C}$ be a centred system of subsets of a certain non-empty set (let us again assume that this set is $\mathbb{N}_+$, this does not affect the argument in any way). Denote $\mathfrak{X}$ the family of all centred systems of subsets of $\mathbb{N}_+$ which contain $\mathscr{C}$. This family is ordered by inclusion. Due to the previous exercise, it is inductive. Since clearly $\cup\mathfrak{C}$ contains $\mathscr{C}$, it follows that $\cup\mathfrak{C}$ belongs to $\mathfrak{X}$ and it forms an upper bound for the chain $\cup\mathfrak{C}$.

Zorn's lemma implies the existence of a maximal element, $\mathscr{D}$, in $\mathfrak{X}$. This $\mathscr{D}$ is a centred system which contains $\mathscr{C}$ and which is not contained in any strictly larger centred system containing $\mathscr{C}$. Since every centred system containing $\mathscr{D}$ will automatically contain $\mathscr{C}$ as well, the latter statement can be cut down to: $\mathscr{D}$ is a centred system which contains $\mathscr{C}$ and which is not contained in any strictly larger centred system. In other words, $\mathscr{D}$ is a maximal centred system containing $\mathscr{C}$, as required.   ▷

Let us reformulate this result as follows.

B.11.4   *Every Centred System Is Contained in an Ultrafilter.*

B.11.5   Consider the following collection of subsets of $\mathbb{N}_+$:

$$\mathscr{C} = \{\mathbb{N}_+ \setminus \{n\}: n \in \mathbb{N}_+\}.$$

In other words, $\mathscr{C}$ consists of all complements to singletons. This system is clearly centred, and its intersection is empty. According to the previous result, there is an ultrafilter $\mathcal{U}$ containing the system $\mathscr{C}$. One has:

$$\cap\mathcal{U} \subseteq \cap\mathscr{C} = \emptyset.$$

Thus, $\mathcal{U}$ is a non-principal (free) ultrafilter.

In fact, one can show that $\mathbb{N}_+$ supports $2^{\mathfrak{c}}$ pairwise different free ultrafilters. They are extremely numerous.

B.11.6

Solution to Exercise B.11.2 Let $n \in \mathbb{N}_+$ and

$$A_1, A_2, \ldots, A_n \in \bigcup \mathfrak{C}$$

be arbitrary elements of $\mathfrak{C}$. For every $i = 1, 2, \ldots, n$ fix a centred system C_i which belongs to the family $\mathfrak{C}$ and contains A_i:

$$A_i \in C_i \in \mathfrak{C}.$$

Elements of the collection C_i, $i = 1, 2, 3, \ldots, n$ of centred systems are pairwise comparable by inclusion between themselves because $\mathfrak{C}$ is totally ordered by inclusion. Every finite totally ordered set has the largest element: for some $i_0 = 1, 2, \ldots, n$ one has

$$\forall i = 1, 2, \ldots, n, \quad C_i \subseteq C_{i_0}.$$

This means in particular that

$$\forall i = 1, 2, \ldots, n, \quad A_i \in C_{i_0}.$$

Since C_{i_0} is a centred system, one concludes

$$A_1 \cap A_2 \cap \ldots \cap A_n \neq \emptyset,$$

as required.

B.12   Ultralimits. Now let us re-examine the existence of our conjunctural limits making sense for every bounded sequence. Notice first that if (ε_n) is a binary sequence, then, according to the initial way we defined ultrafilters in B.8,

$$\lim \varepsilon_n = 1 \iff \{n : \varepsilon_n = 1\} \in \mathcal{U}. \tag{B.4}$$

This definition depends on the choice of an ultrafilter, and for this reason, the corresponding limit is called the *ultralimit along the ultrafilter $\mathcal{U}$*. It is denoted

$$\lim_{n \to \mathcal{U}} \varepsilon_n.$$

(This notation actually makes good sense, as we will see later.)

This definition in the above form will clearly not work in a general case (simply because it is possible that all the terms of a convergent sequence are different from the limit). However, it extends readily to sequences taking *finitely many* distinct values (*Exercise*). How to extend this definition to an arbitrary bounded sequence? In fact, such an extension is quite natural and moreover unique.

B.12.1

Exercise Let $\mathcal{U}$ be an ultrafilter on natural numbers. By approximating a given bounded sequence (x_n) with sequences taking finitely many different values in the ℓ^∞ norm, extend the definition of the ultralimit to all bounded sequences in such a way that it respects all of our required axioms. Show that this extension is well defined (does not depend on an approximation).

B.12.2

Exercise Conclude that, given an ultrafilter $\mathcal{U}$ on the natural numbers, there is a unique way to define the ultralimit along $\mathcal{U}$ on all bounded sequences which satisfies the properties listed in (B.2) as well as the property in Eq. (B.4).

B.12.3

Exercise Conclude that there is a natural correspondence between the ultrafilters on the set of natural numbers and the limits on bounded sequences satisfying the properties stated in (B.2).

At the same time, the definition of an ultralimit based on approximations is not very convenient. We are going to reformulate it now.

B.13 **A Workable Definition of an Ultralimit**. The right definition in a usable form is obtained by adjusting in an obvious way the classical concept of a convergent sequence $x_n \to x$:

$$\forall \varepsilon > 0, \ \exists N, \ \forall n \geq N, \ |x - x_n| < \varepsilon.$$

This means that, given an $\varepsilon > 0$, for "most" values of n the point x_n is within ε from the limit. In this context, "most" means that the set of such n is cofinite.

If $\mathcal{U}$ is an ultrafilter, "most" values of n means that the set of n with this property belongs to $\mathcal{U}$. So we say that a sequence (x_n) of real numbers converges to a real number x along the ultrafilter $\mathcal{U}$,

$$x = \lim_{n \to \mathcal{U}} x_n,$$

if

$$\forall \varepsilon > 0, \ \exists A \in \mathcal{U}, \ \forall n \in A, \ |x - x_n| < \varepsilon.$$

Since every superset of such an A also belongs to $\mathcal{U}$, an equivalent statement is:

$$\forall \varepsilon > 0, \ \{n \in \mathbb{N}_+ : |x - x_n| < \varepsilon\} \in \mathcal{U}.$$

B.13.1

Exercise Show that the definition of the ultralimit given in this subsection is equivalent to the definition in the language of approximations in Exercise B.12.1.

B.13.2

Example If $\mathcal{U} = (n_0)$ is a trivial ultrafilter generated by the natural number n_0, then the statement

$$\lim_{n \to \mathcal{U}} x_n = x$$

means that for every $\varepsilon > 0$,

$$\{n \in \mathbb{N}_+ : |x - x_n| < \varepsilon\} \ni n_0,$$

that is, simply

$$\forall \varepsilon > 0, \ |x - x_{n_0}| < \varepsilon,$$

which means

$$x = x_{n_0}.$$

This is an "uninteresting" case of an ultralimit. The interesting cases correspond to free ultrafilters.

B.13.3

Exercise Show that if $\lim_{n \to \infty} x_n = x$ (in a usual sense), then for every free ultrafilter $\mathcal{U}$ on $\mathbb{N}_+$,

$$\lim_{n \to \mathcal{U}} x_n = x.$$

B.13.4

Exercise Show that, if one has

$$\lim_{n \to \mathcal{U}} x_n = x$$

for *every* free ultrafilter $\mathcal{U}$ on $\mathbb{N}_+$, then

$$\lim_{n\to\infty} x_n = x.$$

Thus, the familiar symbol ∞ essentially means all the free ultrafilters lumped together, and the existence of a limit in the classical sense signifies that all the free ultrafilters on $\mathbb{N}_+$ agree between themselves on what the value of this limit should be. If there is no such agreement, then the limit in the classical sense does not exist, and the sequence (x_n) is divergent. However, every free ultrafilter still gives its own interpretation of what the limit of the sequence is.

B.13.5

Example Recall the alternating sequence

$$\varepsilon_n = (-1)^{n+1}$$

of zeros and ones as in Eq. (B.1) Every ultrafilter $\mathcal{U}$ on $\mathbb{N}$ either contains the set of odd numbers, or the set of even numbers. In the former case, the ultralimit of our sequence is 0, in the latter case, 1. Because of this, the ultrafilters "disagree" between themselves on what the limit of the sequence should be. Consequently, $\lim_{n\to\infty} \varepsilon_n$ does not exist and the sequence is divergent in the classical sense.

B.13.6

Theorem *Every bounded sequence of real numbers has an ultralimit along every ultrafilter on $\mathbb{N}_+$.*

◁ Let (x_n) be a bounded sequence of real numbers, which we, without loss in generality, will assume to belong to the interval $[0, 1]$. Let $\mathcal{U}$ be an ultrafilter on $\mathbb{N}_+$. The proof closely resembles the proof of the Heine–Borel theorem about compactness of the closed unit interval, but is actually simpler.

Subdivide the interval $I_0 = [0, 1]$ into two subintervals of length half, $[0, 1/2]$ and $[1/2, 1]$. Set

$$A = \{n \in \mathbb{N}_+ : x_n \in [0, 1/2]\}.$$

Either A belongs to $\mathcal{U}$, or else its complement, A^c, does. In the latter case, since

$$\{n : x_n \in [1/2, 1]\} \supseteq A^c,$$

the set $\{n : x_n \in [1/2, 1]\}$ is in $\mathcal{U}$ as well. We conclude: at least one of the two intervals of length $1/2$, denote it I_1, has the property

$$\{n \in \mathbb{N}_+ : x_n \in I_1\} \in \mathcal{U}.$$

Continue dividing the intervals in two and selecting one of them. At the end, we will have selected a nested sequence of closed intervals I_k of length 2^{-k}, $k = 0, 1, 2, 3, \ldots$, with the property that for all k,

$$\{n : x_n \in I_k\} \in \mathcal{U}.$$

According to the Cantor Intersection Theorem, there is $c \in [0, 1]$ so that

$$\bigcap_{k=0}^{\infty} I_k = \{c\}.$$

We claim this c is the ultralimit of (x_n) along $\mathcal{U}$. Indeed, let $\varepsilon > 0$ be arbitrary. For a sufficiently large k, the interval I_k is contained in the interval $(c - \varepsilon, c + \varepsilon)$ (indeed, enough to take $k = -\log_2 \varepsilon + 1$). Now one has

$$\{n : c - \varepsilon < x_n < x + \varepsilon\} \supseteq \{n : x_n \in I_k\} \ni \mathcal{U},$$

from where one concludes that the former set is also in $\mathcal{U}$.   ▷

B.13.7

Solution to Exercise B.13.3 We will verify the definition of an ultralimit. Let $\varepsilon > 0$. For some N and every $n \geq N$, one has $|x - x_n| < \varepsilon$. Let us write

$$\mathbb{N} = \{1\} \cup \{2\} \cup \ldots \cup \{N - 1\} \cup \{N, N + 1, N + 2, \ldots\}.$$

One of these sets must belong to $\mathcal{U}$. If it were one of the singletons, $\{i\}$, then every other element $A \in \mathcal{U}$ must meet $\{i\}$, therefore contain i, so $\mathcal{U}$ would be a principal ultrafilter generated by i. One concludes:

$$\{N, N + 1, N + 2, \ldots\} \in \mathcal{U},$$

and so

$$\{n : |x - x_n| < \varepsilon\} \in \mathcal{U},$$

because the set on the left hand side is a superset of $\{N, N + 1, N + 2, \ldots\}$.

B.13.8

Solution to Exercise B.13.4 We have just seen that if $x_n \to x$ in the classical sense, then we have convergence $x_n \overset{\mathcal{U}}{\to} x$ along every free ultrafilter $\mathcal{U}$. This means that the only case to eliminate is where $x_n \overset{\mathcal{U}}{\to} x$ along every free ultrafilter $\mathcal{U}$, yet at the same time $x_n \nrightarrow x$ in the classical sense. Assume that x_n does not converge to x. Then for some $\varepsilon > 0$ and every N there is $n = n(N) \geq N$ with $d(x, x_n) \geq \varepsilon$. The set $I = \{n(N): N \in \mathbb{N}\}$ is infinite. There is a free ultrafilter $\mathcal{U}$ on $\mathbb{N}_+$ containing I, namely an ultrafilter containing the centred system $\{I \setminus \{n\}: n = 1, 2, 3, \ldots\}$. Since $I \in \mathcal{U}$,

$$\{n: d(x, x_n) < \varepsilon\} \notin \mathcal{U},$$

because the set above is disjoint from I. We conclude: $x_n \overset{\mathcal{U}}{\not\to} x$.

B.14   Ultralimits in Metric Spaces. So far, we have only considered ultralimits of sequences of real numbers. Of course the definition extends readily to an arbitrary metric space, as follows. A sequence (x_n) of elements of a metric space (X, d) converges to a point x along an ultrafilter $\mathcal{U}$ on the set of positive natural numbers if for every $\varepsilon > 0$ the set

$$\{n \in \mathbb{N}_+: d(x, x_n) < \varepsilon\}$$

belongs to $\mathcal{U}$.

B.14.1

Exercise Prove that a sequence (x_n) of elements of a metric space X can have at most one ultralimit along a given ultrafilter. That is, let $\mathcal{U}$ be an ultrafilter on $\mathbb{N}_+$. Prove that the ultralimit $\lim_{n \to \mathcal{U}} x_n$, if it exists, is unique.

B.14.2   At the same time, let us note that in an arbitrary metric space, not every bounded sequence of elements needs to have an ultralimit. For instance, if X is a metric space with a 0-1 metric and (x_n) is a sequence of pairwise different elements of X, then for every free ultrafilter $\mathcal{U}$ on $\mathbb{N}_+$, the ultralimit $\lim_{n \to \mathcal{U}} x_n$ does not exist.
    Indeed, let $x \in X$ be arbitrary. The set

$$\{n \in \mathbb{N}_+: d(x, x_n) < 1\}$$

contains at most one element (in the case where $x = x_n$ for some n) and so does not belong to $\mathcal{U}$. We conclude: x_n does not converge to x along $\mathcal{U}$. Since this argument applies to every point $x \in X$, the sequence (x_n) does not have an ultralimit.

B.14.3   Since not every bounded sequence in a metric space X needs to have an ultralimit, a natural question to ask is, can be enlarge X in such a way that every sequence has an ultralimit, much in the same way as we can form a completion of X so that every Cauchy sequence is assigned a limit in it?
    The following result provides a negative answer to this question.

B.14.4

Theorem *For a metric space X, the following conditions are equivalent.*

1. *X is compact.*
2. *Every sequence of elements in X has an ultralimit along every free ultrafilter $\mathcal{U}$ on $\mathbb{N}_+$.*
3. *There exists a free ultrafilter $\mathcal{U}$ on $\mathbb{N}_+$ with the property that every sequence of elements in X has an ultralimit along $\mathcal{U}$.*

◁ (a) ⇒ (b): very much the same proof as for the interval. For every $\varepsilon > 0$, choose a finite cover of X with open ε-balls. One of these balls, say B, has the property

$$\{n : x_n \in B\} \in \mathcal{U}.$$

Proceeding recursively, we obtain a centred sequence of open balls B_i of radius converging to zero satisfying

$$A_i = \{n : x_n \in B_i\} \in \mathcal{U}.$$

The ball centers form a Cauchy sequence and so converge to some limit, x. Every neighbourhood V of x contains some B_i for i large enough, and so the set

$$\{n : x_n \in V\}$$

contains A_i and so belongs to $\mathcal{U}$.

(b)⇒(c): trivial, given that free ultrafilters on $\mathbb{N}$ exist. Now just pick any one of them.

(c)⇒(a): by contraposition. If X is not compact, then either it is not totally bounded (in which case there is a sequence of points at pairwise distances $\geq \varepsilon_0 > 0$ from each other, which has no ultralimit due to the argument in B.14.2), or non-complete. In the latter case, select a Cauchy sequence (x_n) without a limit. We will show that it does not admit an ultralimit along $\mathcal{U}$ either. Assume x is such an ultralimit. For every $\varepsilon > 0$, the set $\{n : d(x_n, x) < \varepsilon\}$ is in $\mathcal{U}$, so non-empty, and using this, one can recursively select a subsequence of x_n converging to x in the usual sense. But (x_n) is a Cauchy sequence, so this would mean $x_n \to x$.   ▷

B.14.5

Exercise Deduce that a metric space X metrically embeds into a space $\hat{X}$ in which every bounded sequence has an ultralimit if and only if every ball in X is totally bounded. In this case, $\hat{X}$ is the metric completion of X, and every closed ball in X is compact.

This is of course a rather restrictive condition, which, for instance, is failed by any infinite-dimensional normed space.

B.14.6

Solution to Exercise B.14.1 Suppose $x, y \in X$ and

$$\lim_{n \to \mathcal{U}} x_n = x \text{ and } \lim_{n \to \mathcal{U}} x_n = y.$$

For every $\varepsilon > 0$, both sets

$$\{n \in \mathbb{N}_+ : d(x, x_n) < \varepsilon\} \text{ and } \{n \in \mathbb{N}_+ : d(y, x_n) < \varepsilon\}$$

belong to the ultrafilter $\mathcal{U}$, and so does their intersection. As a consequence, this intersection is non-empty, and one can find $n \in \mathbb{N}$ with

$$d(x, x_n) < \varepsilon \text{ and } d(y, x_n) < \varepsilon.$$

By the triangle inequality,

$$d(x, y) < 2\varepsilon,$$

and since this holds for every $\varepsilon > 0$, we conclude: $x = y$.

B.15 **Ultraproducts.** In view of (B.14.5), we cannot hope to attach a "virtual" ultralimit to every bounded sequence of points of a metric space, X. If the space is not totally bounded, some sequences are destined to diverge along some ultrafilters in every *metric* extension of X. Still, we can assign to every such sequence an ideal new point, which will be the limit (hence, ultralimit) in case of a Cauchy sequence, and otherwise will register the asymptotic behaviour of the sequence with regard to the given ultrafilter. Now we will briefly examine this important construction.

B.15.1 *Space of Bounded Sequences in a Metric Space.* Let $X = (X, d_X)$ be a metric space. Fix an ultrafilter $\mathcal{U}$ on the integers. Since we are going to assign an "ideal" point to every bounded sequence, the right place to start will be the set $\ell^\infty(\mathbb{N}_+; X)$ of all bounded sequences with elements in X. (For example, the space ℓ^∞ is, in this notation, $\ell^\infty(\mathbb{N}_+; \mathbb{R})$.) We would formally identify the space of "ideal" points with $\ell^\infty(\mathbb{N}_+; X)$, or rather with its quotient space under a suitable equivalence relation—just like we do when we construct the completion of a metric space beginning with the set of all Cauchy sequences.

Observe that the set $\ell^\infty(\mathbb{N}_+; X)$ admits a structure of a metric space with regard to the ℓ^∞-distance:

$$d_\infty(x, y) = \sup_{n \in \mathbb{N}_+} d_X(x_n, y_n).$$

B.15.2

Exercise Show that the metric space $\ell^\infty(\mathbb{N}_+; X)$ is complete if and only if X is complete.

B.15.3 It is quite natural to make two bounded sequences x and y share the same "ideal point" if the distance between the corresponding terms of those sequences converges to zero *along the ultrafilter* $\mathcal{U}$:

$$x \overset{\mathcal{U}}{\sim} y \iff \lim_{n \to \mathcal{U}} d_X(x_n, y_n) = 0.$$

The resulting equivalence relation $\overset{\mathcal{U}}{\sim}$ on $\ell^\infty(\mathbb{N}_+; X)$ agrees with the metric in the sense that the quotient metric on the quotient set is well defined. For an $x \in \ell^\infty(\mathbb{N}_+; X)$ denote

$$[x]_{\mathcal{U}} = \{y \in X : x \overset{\mathcal{U}}{\sim} y\}$$

the corresponding equivalence class.

B.15.4

Exercise Show that the rule

$$d([x], [y]) = \inf\{d_\infty(x', y') : x' \in [x]_{\mathcal{U}}, y' \in [y]_{\mathcal{U}}\}$$

defines a metric on the quotient set $X / \overset{\mathcal{U}}{\sim}$, and that this quotient metric can be alternatively described by

$$d([x], [y]) = \lim_{n \to \mathcal{U}} d_X(x_n, y_n).$$

B.15.5   *Metric Ultrapowers.* The quotient metric space $\ell^\infty(\mathbb{N}_+; X) / \overset{\mathcal{U}}{\sim}$ is called the *metric ultrapower* of X with regard to the ultrafilter $\mathcal{U}$, and denoted $X_{\mathcal{U}}^{\mathbb{N}}$.

B.15.6

Exercise Show that the original metric space X canonically isometrically embeds inside of its metric ultrapower under the *diagonal embedding*

$$X \ni x \mapsto [(x, x, x, \ldots)]_{\mathcal{U}} \in X_{\mathcal{U}}^{\mathbb{N}},$$

no matter what the ultrafilter $\mathcal{U}$ is. In particular, the metric ultrapower of a non-trivial space is itself non-trivial.

B.15.7

Exercise Show that, if $\mathcal{U} = (n)$ is a principal ultrafilter, the metric ultrapower $X_{\mathcal{U}}^{\mathbb{N}}$ is canonically isometric to X.

B.15.8

Exercise Let $\mathcal{U}$ be a non-principal ultrafilter, and suppose that X is a *separable* metric space. Show that the metric ultrapower $X_{\mathcal{U}}^{\mathbb{N}}$ is isometric to X (not necessarily in a canonical way!) if and only if X is compact.

Before establishing a couple of other properties of the ultrapower, let us extend the definition to the case where terms of sequences come from *possibly different* metric spaces: for every n,

$$x_n \in X_n,$$

where (X_n), $n = 1, 2, \ldots$, are metric spaces which may or may not differ from each other.

B.15.9 *Pointed Metric Spaces.* In this situation, how do we define a bounded sequence $x = (x_n)$? The notion of boundedness becomes purely relative: one can say that a sequence y is bounded with regard to another sequence, x, if the values $d_{X_n}(x_n, y_n)$, $n \in \mathbb{N}$, form a bounded set. However, there is no absolute notion of boundedness. For this reason, it is necessary to select a sequence $x^* = (x_n^*)$, $x_n^* \in X_n$, thus fixing a class of sequences bounded with regard to x^*. In other words, we are dealing with a family of *pointed metric spaces*, (X_n, x_n^*).

For example, in cases of considerable interest, where X_n are normed spaces or metric groups, the selected points are usually zero and the neutral element, respectively. If all spaces $X_n = X$ are the same, in order to obtain the usual notion of (absolute) boundedness of a sequence, one chooses any constant sequence $(x^*, x^*, \ldots)$, all of them giving the same result. If the spaces X_n have a uniformly bounded diameter, the choice of a distinguished sequence does not matter.

B.15.10 *Ultraproduct of Pointed Metric Spaces.* Now it is clear how to reformulate the definitions. The metric space consisting of all relatively bounded sequences with regard to x^* is the ℓ^∞-type sum of pointed spaces (X_n, x_n^*), denoted

$$\oplus_n^{\ell^\infty} (X_n, x_n^*).$$

In cases like the above where the choice of distinguished points is clear or does not matter, these points are suppressed.

The equivalence relation and the distance on the quotient metric space are defined in exactly the same fashion as above. The resulting metric space

$$\oplus_n^{\ell^\infty} (X_n, x_n^*) / \overset{\mathcal{U}}{\sim}$$

is called the *metric ultraproduct* of (pointed) metric spaces (x_n, x_n^*) modulo the ultrafilter $\mathcal{U}$, and denoted

$$\left(\prod_n (X_n, x_n^*) \right)_{\mathcal{U}} .$$

B.15.11

Exercise Assuming that $\mathcal{U}$ is a non-principal ultrafilter, show that there is a canonical isometry between the metric ultraproducts

$$\left(\prod_n (X_n, x_n^*)\right)_{\mathcal{U}} \quad \text{and} \quad \left(\prod_n (\widehat{X}_n, x_n^*)\right)_{\mathcal{U}} ,$$

where $\widehat{X}_n$ denotes the completion of the metric space X_n.

◁ *Hint:* observe that the ultraproduct on the left admits a canonical isometric embedding into the one on the right.   ▷

B.15.12

Exercise Given a Cauchy sequence in the metric ultraproduct of spaces X_n, show that it is the image under the quotient map of some Cauchy sequence in the ℓ^∞-type sum of the spaces X_n.

◁ *Hint:* use the following equivalent definition: a sequence (a_n) is Cauchy if and only if for every $\varepsilon > 0$ there is N such that the open ball $B_\varepsilon(x_N)$ contains all x_n with $n \geq N$.   ▷

The following stands in contrast to Exercise B.15.2.

B.15.13

Exercise Deduce from the previous two exercises that the metric ultraproduct of a family of arbitrary metric spaces, formed with regard to a non-principal ultrafilter, is always a complete metric space.

Recall that the covering number $N(X, \varepsilon)$, where $\varepsilon > 0$, of a metric space X is the smallest size of an ε-net for X.

B.15.14

Exercise (∗) Prove that if the metric ultraproduct X of a family (X_n) of separable metric spaces is separable, then it is compact. Moreover, in this case for every $\varepsilon > 0$ the sequence of covering numbers $N(X_n, \varepsilon)$ is essentially bounded, that is, uniformly bounded from above for all n belonging to some element $A = A(\varepsilon)$ of the ultrafilter.

Otherwise, X has density character continuum.

B.16   *The Space of Ultrafilters.* In conclusion, we want to give a clear literal meaning to the symbol

$$\lim_{n \to \mathcal{U}} x_n,$$

used for an ultralimit along an ultrafilter $\mathcal{U}$. In fact, an ultralimit of a sequence can be interpreted as a limit in the usual sense.

B.16.1 Denote by $\beta\mathbb{N}$ the set of all ultrafilters on the set $\mathbb{N}$ of natural numbers. Identify $\mathbb{N}$ with a subset of $\beta\mathbb{N}$ by assigning to each natural number n the corresponding principal ultrafilter, $(n) = \{A \subseteq \mathbb{N}: A \ni n\}$:

$$\mathbb{N} \hookrightarrow \beta\mathbb{N}, \quad \mathbb{N} \ni n \mapsto (n) \in \beta\mathbb{N}.$$

This mapping is clearly one-to-one, and in this sense we will often refer to $\mathbb{N}$ as a subset of $\beta\mathbb{N}$.

B.16.2 For every $A \subseteq \mathbb{N}$, denote by $\tilde{A}$ the family of all ultrafilters $\xi \in \beta\mathbb{N}$ containing A as an element:

$$\tilde{A} = \{\mathcal{U} \in \beta\mathbb{N}: A \in \mathcal{U}\}.$$

B.16.3

Exercise Prove that the collection

$$\{\tilde{A}: A \subseteq \mathbb{N}\}$$

forms a base for a topology on $\beta\mathbb{N}$.

B.16.4

Exercise Prove that the above described topology on $\beta\mathbb{N}$ induces the discrete topology on $\mathbb{N}$ as on a topological subspace.

B.16.5

Exercise Prove that the topology as above on $\beta\mathbb{N}$ is Hausdorff.

B.16.6

Exercise Prove that $\beta\mathbb{N}$ with the above topology is compact.

B.16.7

Exercise Prove that $\mathbb{N}$ forms an *everywhere dense subset* in $\beta\mathbb{N}$, that is, the closure of $\mathbb{N}$ in $\beta\mathbb{N}$ is the entire space $\beta\mathbb{N}$.

B.16.8

Exercise Let X be an arbitrary compact space, and let $f: \mathbb{N} \to X$ be an arbitrary mapping. Prove that f extends to a continuous mapping $\tilde{f}: \beta X \to X$.

B.16.9

Exercise Prove that a continuous mapping $\tilde{f}$ as above is unique for any given f, assuming the space X is compact Hausdorff. In other words, if $g\colon \beta\mathbb{N} \to X$ is a continuous mapping with $g|_{\mathbb{N}} = f$, then $g = \tilde{f}$.

B.16.10   The compact space $\beta\mathbb{N}$ as above is called the *Stone–Čech compactification* (or else *universal compactification*) of the discrete space $\mathbb{N}$.

B.16.11

Exercise Let $\mathcal{U} \in \beta\mathbb{N}$ be an ultrafilter, and let $x = (x_n)$ be a sequence of points in a compact metric space X, in other words, a function $x\colon \mathbb{N} \to X$. Prove that the ultralimit of (x_n) along $\mathcal{U}$ is exactly the value of the unique continuous extension $\tilde{x}$ at $\mathcal{U}$, that is, the usual classical limit of the function x as $n \to \mathcal{U}$:

$$\lim_{n \to \mathcal{U}} x_n = \tilde{x}(\mathcal{U}) = \lim_{n \to \mathcal{U}} x.$$

B.16.12

Exercise Generalize the above as follows: suppose $\mathcal{U}$ is an ultrafilter on $\mathbb{N}$, and let (x_n) be a sequence of points in a metric space X. Prove that the following are equivalent:

1. The sequence (x_n) converges along the ultrafilter $\mathcal{U}$ to some point $x \in X$.
2. The function $n \mapsto x_n$ has a classical limit $x \in X$ as n approaches the point $\mathcal{U}$ in the topological space $\beta\mathbb{N}$.

Moreover, if the two limits in (1) and (2) exist, they are equal.

B.16.13

Solution to Exercise B.16.3 Since every ultrafilter $\mathcal{U}$ on $\mathbb{N}$ contains $\mathbb{N}$ and thus $\mathcal{U} \in \tilde{\mathbb{N}}$, one concludes that $\tilde{\mathbb{N}} = \beta\mathbb{N}$ and therefore

$$\cup_{A \subseteq \mathbb{N}} \tilde{A} = \beta\mathbb{N}.$$

Now let $A, B \subseteq \mathbb{N}$, and let $\mathcal{U} \in \tilde{A} \cap \tilde{B}$. Then $\mathcal{U} \ni A$ and $\mathcal{U} \ni B$, therefore $\mathcal{U} \ni A \cap B$. It means that $\mathcal{U} \in \widetilde{A \cap B}$. On the other hand, clearly $\widetilde{A \cap B} \subseteq \tilde{A} \cap \tilde{B}$ because, more generally, if $C \subseteq D \subseteq \mathbb{N}$, then $\tilde{C} \subseteq \tilde{D}$ (un ultrafilter containing C necessarily contains D as well). We have proved that $\widetilde{A \cap B} = \tilde{A} \cap \tilde{B}$, from where the second axiom of a base follows.

B.16.14

Solution to Exercise B.16.4 Here one should notice that $\mathbb{N}$ is naturally identified with a subset of $\beta\mathbb{N}$ through identifying each element $n \in \mathbb{N}$ with the principal (trivial) ultrafilter, (n), generated by n:

$$\mathbb{N} \ni n \mapsto (n) = \{A \subseteq \mathbb{N} : A \ni n\} \in \beta\mathbb{N}.$$

This mapping is clearly one-to-one, and so we will often think of $\mathbb{N}$ as a subset of $\beta\mathbb{N}$.

To establish the claim, it is enough to show that every singleton in $\mathbb{N}$ is open in the induced topology. Let $n \in \mathbb{N}$. Then it is easy to see that

$$\widetilde{\{n\}} = \{(n)\},$$

because if an ultrafilter $\mathcal{U}$ contains the set $\{n\}$, it must coincide with the trivial ultrafilter (n). Since the singleton $\{(n)\}$ is open in $\beta\mathbb{N}$, it is also open in $\mathbb{N}$. ▷

B.16.15

Solution to Exercise B.16.5 Let $\mathcal{U}, \zeta \in \beta\mathbb{N}$, and let $\mathcal{U} \neq \zeta$. What is means, is this: there is an $A \subseteq \mathbb{N}$ such that A belongs to one ultrafilter (say $\mathcal{U}$) and not the other (that is, $A \notin \zeta$). A major property of ultrafilters implies then that $\mathbb{N} \setminus A \in \zeta$. The sets $\tilde{A}$ and $\widetilde{\mathbb{N} \setminus A}$ are both open in $\beta\mathbb{N}$, contain $\mathcal{U}$ and ζ respectively, and are disjoint: indeed, assuming $\kappa \in \tilde{A} \cap \widetilde{\mathbb{N} \setminus A}$ would mean that the ultrafilter κ contains both A and $\mathbb{N} \setminus A$, which is impossible.

B.16.16

Solution to Exercise B.16.6 Let γ be an arbitrary open cover of $\beta\mathbb{N}$. Then

$$\beta := \{\tilde{A} : \text{for some } V \in \gamma, \tilde{A} \subseteq V\}$$

is also an open cover of $\beta\mathbb{N}$: indeed, if $\mathcal{U} \in \beta\mathbb{N}$, then for some $V \in \gamma$ one has $\mathcal{U} \in V$, and by the very definition of the topology on $\beta\mathbb{N}$, there is an $A \subseteq \mathbb{N}$ with $\mathcal{U} \in \tilde{A} \subseteq V$. Notice that it is now enough to select a finite subcover, say β_1, of β: this done, we will then replace each $\tilde{A} \in \beta_1$ with an arbitrary $V \in \gamma$ such that $V \supseteq \tilde{A}$, and thus we will get a finite subcover of γ. From now on, we can forget of γ and concentrate on β alone, and the cover β consists of basic open sets.

Consider the system of subsets of $\mathbb{N}$,

$$\delta := \{A \subseteq \mathbb{N} : \tilde{A} \in \beta\}.$$

This is clearly a cover of $\mathbb{N}$: $\cup\delta = \mathbb{N}$. If we assume that δ contains no finite subcover, it is the same as to assume that the system of complements,

$$\{\mathbb{N} \setminus A : A \in \delta\},$$

is centred. Being centred, it is contained in an ultrafilter, say $\mathcal{U}$. One has

$$\forall A \in \delta, \ \mathbb{N} \setminus A \in \mathcal{U},$$

and consequently,

$$\forall A \in \delta, \ A \notin \mathcal{U},$$

which can be rewritten as

$$\forall A \in \delta, \ \mathcal{U} \notin \tilde{A},$$

that is,

$$\mathcal{U} \notin \cup \beta,$$

a contradiction. One concludes: there are finitely many elements $A_1, A_2, \ldots, A_n \in \delta$ with

$$A_1 \cup A_2 \cup \cdots \cup A_n = \mathbb{N}.$$

By force of a familiar property of ultrafilters, if $\mathcal{U}$ is an arbitrary ultrafilter on $\mathbb{N}$, it must contain at least one of the sets A_i, $i = 1, \ldots, n$, or, equivalently, $\mathcal{U} \in \tilde{A}_i$ for some $i = 1, 2, \ldots, n$. It follows that

$$\widetilde{A_1} \cup \widetilde{A_2} \cup \cdots \cup \widetilde{A_n} = \beta \mathbb{N},$$

which of course finishes the proof by supplying the desired finite subcover β_1 of β.

B.16.17

Solution to Exercise B.16.7 It is enough to find an element of $\mathbb{N}$ in an arbitrary non-empty open subset of $\beta \mathbb{N}$, say U. Since U is the union of basic open subsets, for some $A \subseteq \mathbb{N}$, $A \neq \emptyset$, one has $\tilde{A} \subseteq U$. Let $a \in A$ be arbitrary. Then $A \in (a)$, that is, $(a) \in \tilde{A} \subseteq U$, as required.

B.16.18

Solution to Exercise B.16.8 Let $\mathcal{U} \in \beta \mathbb{N}$ be arbitrary. Since the space X is compact, there is a limit, $\lim_{n \to \mathcal{U}} f(x) \in X$. Denote this limit $\tilde{f}(\mathcal{U})$. It remains to prove the continuity of $\tilde{f}$. Let $U \subseteq X$ be open, and let $\tilde{f}(\mathcal{U}) \in U$ for some $\mathcal{U} \in \beta \mathbb{N}$. We want to find a neighbourhood V of $\mathcal{U}$ in $\beta \mathbb{N}$ such that $\tilde{f}(V) \subseteq U$.

Every compact space (T_1 or not) is regular: this is easily proved using an argument involving finite subcovers of covers of closed sets with open neighbourhoods.

Therefore, one can find an open set, W, in X with

$$\tilde{f}(\mathcal{U}) \in W \subseteq \mathrm{cl}\, W \subseteq U.$$

The condition $\tilde{f}(\mathcal{U}) \in W$, that is, $\lim_{n \to \mathcal{U}} f(x) \in W$, implies that $f^{-1}(W) \in \mathcal{U}$. Set $A := f^{-1}(W)$. It is a non-empty subset of $\mathbb{N}$, and therefore $V := \bar{A}$ forms an open neighbourhood of $\mathcal{U}$ in $\beta\mathbb{N}$. If now $\zeta \in V$, that is, $\zeta \ni A = f^{-1}(W)$, then every neighbourhood of the limit, $\tilde{f}(\zeta)$, in X meets W, meaning that $\tilde{f}(\zeta) \in \mathrm{cl}\, W$ and consequently $\tilde{f}(\zeta) \in U$, as required. We conclude: $\tilde{f}(V) \subseteq W$. The continuity of the mapping

$$\tilde{f}: \beta\mathbb{N} \to X$$

is thus established.

B.16.19

Solution to Exercise B.16.9 This follows at once from a much more general assertion, making no use of compactness whatsoever: if $f, g : X \to Y$ are continuous mappings between two topological spaces, where $Y \in T_2$, and if $Z \subseteq X$ is everywhere dense in X, and if $f|_Z = g|_Z$, then $f = g$.

Indeed, assume that for some $x \in X$ one has

$$f(x) \neq g(x).$$

Find disjoint open neighbourhoods V and U in Y of $f(x)$ and $g(x)$, respectively. Their preimages, $f^{-1}(V)$ and $g^{-1}(U)$, form open neighbourhoods of x in X, and so does their intersection. Since Z is everywhere dense in X, there is a $z \in Z$ such that $z \in f^{-1}(V) \cap g^{-1}(U)$. Now one has

$$V \ni f(z) = g(z) \in U,$$

a contradiction since $V \cap U = \emptyset$.

Without the assumption that X be Hausdorff, the statement is no longer true. A simple example is this: let $X = \{0, 1\}$ be a two-element set with indiscrete topology (that is, only $\emptyset$ and all of X are open). Clearly, such an X is compact. Define a map $f : \mathbb{N} \to X$ as the constant map, sending each $n \in \mathbb{N}$ to 0. Since an arbitrary map from any topological space to an indiscrete space is always continuous, the map f admits more than one extension to a map to X, and all of them are continuous. (For example, one can send all elements of the remainder $\beta\mathbb{N} \setminus \mathbb{N}$ to 1, or else to 0.)

Thanks to Tullio G. Ceccherini-Silberstein and Michel Coornaert for a number of remarks on this appendix.

Bibliography

1. H. Abels, An example of a finitely presented solvable group, in *Homological Group Theory (Proc. Sympos., Durham, 1977)*. London Mathematical Society Lecture Note Series, vol. 36 (Cambridge University Press, Cambridge, 1979), pp. 205–211

2. H. Ando, U. Haagerup, C. Winslow, Ultraproducts, QWEP von Neumann Algebras, and the Effros-Marechal Topology (2013). Preprint, available at arXiv:1306.0460v1

3. P. Ara, K.C. O'Meara, F. Perera, Stable finiteness of group rings in arbitrary characteristic. Adv. Math. **170**(2), 224–238 (2002)

4. E. Artin, Über die zerlegung definiter functionem in quadrate. Abh. Math. Sem. Univ. Hamburg **5**, 85–89 (1927)

5. G. Arzhantseva, Asymptotic approximations of finitely generated groups (2012). Preprint, available at http://www.mat.univie.ac.at/~arjantseva/Abs/asymptotic.pdf

6. G. Arzhantseva, L. Păunescu, Linear sofic groups and algebras (2012). Preprint, available at arXiv:1212.6780

7. G. Arzhantseva, L. Paunescu, Almost commuting permutations are near commuting permutations (October 2014). arXiv:1410.2626

8. J. Bannon, A non-residually solvable hyperlinear one-relator group. Proc. Am. Math. Soc. **139**(4), 1409–1410 (2011)

9. J. Bannon, N. Noblett, A note on non-residually finite solvable hyperlinear groups. Involve **3**, 345–347 (2010)

10. G. Baumslag, A non-cyclic one-relator group all of whose finite quotients are cyclic. J. Aust. Math. Soc. Ser. A. Pure Math. Stat. **10**, 497–498 (1969)

11. G. Baumslag, D. Solitar, Some two-generator one-relator non-hopfian groups. Bull. Am. Math. Soc. **68**(3), 199–201 (1962)

12. B. Bekka, P. de la Harpe, A. Valette, *Kazhdan's Property (T)*. New Mathematical Monographs, vol. 11 (Cambridge University Press, Cambridge, 2008)

13. I. Ben Yaacov, A. Berenstein, C.W. Henson, A. Usvyatsov, Model theory for metric structures, in *Model Theory with Applications to Algebra and Analysis*, vol. 2. London Mathematical Society Lecture Note Series, vol. 350 (Cambridge University Press, Cambridge, 2008), pp. 315–427

14. B. Blackadar, Theory of C^*-algebras and von Neumann algebras. Operator algebras and Non-commutative geometry, III, in *Operator Algebras*. Encyclopaedia of Mathematical Sciences, vol. 122 (Springer, Berlin, 2006)

15. L. Bowen, Measure conjugacy invariants for actions of countable sofic groups. J. Am. Math. Soc. **23**(1), 217–245 (2010)

© Springer International Publishing Switzerland 2015

V. Capraro, M. Lupini, *Introduction to Sofic and Hyperlinear Groups and Connes' Embedding Conjecture*, Lecture Notes in Mathematics 2136, DOI 10.1007/978-3-319-19333-5

16. N.P. Brown, Connes' embedding problem and Lance's WEP. Int. Math. Rev. Not. **10**, 501–510 (2004)

17. N.P. Brown, Topological dynamical systems associated to II_1-factors. Adv. Math. **227**(4), 1665–1699 (2011) [With an appendix by Narutaka Ozawa]

18. N.P. Brown, V. Capraro, Groups associated to II_1-factors. J. Funct. Anal. **264**(2), 493–507 (2013)

19. M. Burger, A. Valette, Idempotents in complex group rings: theorems of Zalesskii and Bass revisited. J. Lie Theory **8**(2), 219–228 (1998)

20. V. Capraro, Cross norms on tensor product of universal C*-algebras and Kirchberg property. Mathoverflow question (2012). http://mathoverflow.net/questions/90686/cross-norms-on-tensor-product-of-universal-c-algebras-and-kirchberg-property

21. V. Capraro, Double ultrapower of the hyperfinite II_1-factor. Mathoverflow question (2013). http://mathoverflow.net/questions/133437/double-ultrapower-of-the-hyperfinite-ii-1-factor

22. V. Capraro, T. Fritz, On the axiomatization of convex subsets of a Banach space. Proc. Am. Math. Soc. **141**(6), 2127–2135 (2013)

23. V. Capraro, K. Morrison, Optimal strategies for a game on amenable semigroups. Int. J. Game Theory **42**(4), 917–929 (2013)

24. V. Capraro, L. Păunescu, Product between ultrafilters and applications to Connes' embedding problem. J. Oper. Theory **68**(1), 165–172 (2012)

25. V. Capraro, F. Rădulescu, Cyclic Hilbert spaces and Connes' embedding problem. Compl. Anal. Oper. Theory **7**, 863–872 (2013)

26. V. Capraro, M. Scarsini, Existence of equilibria in countable games: an algebraic approach. Games Econ. Behav. **79**, 163–180 (2013)

27. K. Carlson, E. Cheung, I. Farah, A. Gerhardt-Bourke, B. Hart, L. Mezuman, N. Sequeira, A. Sherman, Omitting types and AF algebras. Arch. Math. Log. **53**(1–2), 157–169 (2014)

28. A. Castilla, Artin-Lang property for analytic manifolds of dimension two. Math. Z. **217**, 5–14 (1994)

29. A. Cayley, On the theory of groups, as depending on the symbolic equation $\theta^n = 1$, Part II. Philos. Mag. Ser. 4 **7**(47), 408–409 (1854)

30. T. Ceccherini-Silberstein, M. Coornaert, Injective linear cellular automata and sofic groups. Isr. J. Math. **161**(1), 1–15 (2007)

31. T. Ceccherini-Silberstein, M. Coornaert, *Cellular Automata and Groups*. Springer Monographs in Mathematics (Springer, Berlin, 2010)

32. T. Ceccherini-Silberstein, M. Coornaert, On sofic monoids. Semigroup Forum **89**(3), 546–570 (2014)

33. T. Ceccherini-Silberstein, M. Coornaert. On surjunctive monoids (September 2014). arXiv:1409.1340

34. C.C. Chang, H.J. Keisler, *Model Theory* (Princeton University Press, Princeton, 1966)

35. A. Chirvasitu, Dedekind complete posets from sheaves on von Neumann algebras (2013). Preprint, available at arXiv:1307.8400v1

36. M.D. Choi, A Schwarz inequality for positive linear maps on C*-algebras. Ill. J. Math. **18**, 565–574 (1974)

37. M.D. Choi, The full C*-algebra of the free group on two generators. Pac. J. Math. **87**, 41–48 (1980)

38. L. Ciobanu, D.F. Holt, S. Rees, Sofic groups: graph products and graphs of groups. Pac. J. Math. **271**(1), 53–64 (2014)

39. P.J. Cohen, The independence of the continuum hypothesis. Proc. Natl. Acad. Sci. USA **50**, 1143–1148 (1963)

40. P.J. Cohen. The independence of the continuum hypothesis, II. Proc. Natl. Acad. Sci. USA **51**, 105–110 (1964)

41. B. Collins, K. Dykema, Linearization of Connes' embedding problem. N. Y. J. Math. **14**, 617–641 (2008)

42. B. Collins, K. Dykema, Free products of sofic groups with amalgamation over monotileably amenable groups. Münster J. Math. **4**, 101–118 (2011)

43. A. Connes, Classification of injective factors. Cases $II_1, II_\infty, III_\lambda, \lambda \neq 1$. Ann. Math. 2 **104**(1), 73–115 (1976)

44. A. Connes, V. Jones, Property T for von Neumann algebras. Bull. Lond. Math. Soc. **17**(1), 57–62 (1985)

45. Y. Cornulier, A sofic group away from amenable groups. Math. Ann. **350**(2), 269–275 (2011)

46. Y. de Cornulier, L. Guyot, W. Pitsch, On the isolated points in the space of groups. J. Algebra **307**(1), 254–277 (2007)

47. Y. de Cornulier, Finitely presentable, non-hopfian groups with kazhdan's property (t) and infinite outer automorphism group. Proc. Am. Math. Soc. **135**(4), 951–959 (2007)

48. C.N. Delzell, A continuous, constructive solution to Hilbert's 17th problem. Invent. Math. **76**(3), 365–384 (1984)

49. J. Dodziuk, P. Linnell, V. Mathai, T. Schick, S. Yates, Approximating L^2-invariants and the Atiyah conjecture. Commun. Pure Appl. Math. **56**(7), 839–873 (2003) [Dedicated to the memory of Jürgen K. Moser]

50. H.A. Dye, On the geometry of projections in certain operator algebras. Ann. Math. 2 **61**, 73–89 (1955)

51. E. Effros, Convergence of closed subsets in a topological space. Proc. Am. Math. Soc. **16**, 929–931 (1965)

52. E. Effros, Global structure in von Neumann algebras. Trans. Am. Math. Soc. **121**, 434–454 (1966)

53. G. Elek, E. Szabó, Sofic groups and direct finiteness. J. Algebra **280**, 426–434 (2004)

54. G. Elek, E. Szabó, Hyperlinearity, essentially free actions and L^2-invariants. the sofic property. Math. Ann. **332**(2), 421–441 (2005)

55. G. Elek, E. Szabó, On sofic groups. J. Group Theory **9**(2), 161–171 (2006)

56. G. Elek, E. Szabó, Sofic representations of amenable groups. Proc. Am. Math. Soc. **139**, 4285–4291 (2011)

57. I. Farah, B. Hart, M. Lupini, L. Robert, A.P. Tikuisis, A. Vignati, W. Winter, Model theory of nuclear C*-algebras (in preparation). Available at http://www.math.yorku.ca/~ifarah/Ftp/ 2014b22-good.pdf

58. I. Farah, B. Hart, D. Sherman, Model theory for operator algebras III: elementary equivalence and II_1 factors. Bull. Lond. Math. Soc. **46**(3), 609–628 (2014)

59. I. Farah, B. Hart, D. Sherman, Model theory for operator algebras I: stability. Bull. Lond. Math. Soc. **45** 825–838 (2013)

60. I. Farah, B. Hart, D. Sherman, Model theory for operator algebras II: model theory. Isr. J. Math. **201**(1), 477–505 (2014)

61. I. Farah, S. Shelah, A dichotomy for the number of ultrapowers. J. Math. Log. **10**(1–2), 45–81 (2010)

62. D. Farenick, V. Paulsen, Operator system quotients of matrix algebras and their tensor products. Math. Scand. **III**, 210–243 (2012)

63. D. Farenick, A.S. Kavruk, V.I. Paulsen, C*-algebras with the weak expectation property and a multivariable analogue of Ando's theorem on the numerical radius. J. Oper. Theory **70**(2), 573–590 (2013)

64. D. Farenick, A.S. Kavruk, V.I. Paulsen, I.G. Todorov. Characterisations of the weak expectation property (2013). Available at arXiv:1307.1055

65. M. Fekete, Uber die verteilung der wurzeln bei gewissen algebraischen gleichungen mit. ganzzahligen koeffizienten. Math. Z. **17**, 228–249 (1923)

66. T. Fritz, Tsirelson's problem and Kirchberg's conjecture. Rev. Math. Phys. **24**(1250012), 67 pp. (2012)

67. W. Fulton, *Algebraic Topology: A First Course* (Springer, New York, 2008)

68. I. Gelfand, M. Neumark, On the imbedding of normed rings into the ring of operators in Hilbert space, in *C*-Algebras: 1943–1993 (San Antonio, TX, 1993)*. Contemporary Mathematics, vol. 167 (American Mathematical Society, Providence, 1994), pp. 2–19 (Corrected reprint of the 1943 original [MR **5**, 147])

69. M. Gerstenhaber, O.S. Rothaus, The solution of sets of equations in groups. Proc. Natl. Acad. Sci. USA **48**, 1531–1533 (1962)
70. L. Glebsky, L.M. Rivera, Sofic groups and profinite topology on free groups. J. Algebra **320**, 3512–3518 (2008)
71. L. Glebsky, L.M. Rivera, Almost solutions of equations in permutations. Taiwan. J. Math. **13**(2A), 493–500 (2009)
72. K. Gödel, *The Consistency of the Continuum Hypothesis*. Annals of Mathematics Studies, vol. 3 (Princeton University Press, Princeton, 1940)
73. I. Goldbring, B. Hart, A computability-theoretic reformulation of the Connes Embedding Problem (2013). Preprint, available at arXiv:1308.2638
74. D. Gondard, P. Ribenoim. Le 17eme probleme de Hilbert pour les matrices. Bull. Sci. Math. **98**, 49–56 (1974)
75. E.I. Gordon, A.M. Vershik, Groups that are locally embeddable in the class of finite groups. St. Petersburg Math. J. **9**, 49–67 (1998)
76. W. Gottschalk, Some general dynamical notions, in *Recent Advances in Topological Dynamics (Proc. Conf. Topological Dynamics, Yale Univ., New Haven, Conn., 1972; in honor of Gustav Arnold Hedlund)*. Lecture Notes in Mathematics, vol. 318 (Springer, Berlin, 1973), pp. 120–125
77. F.P. Greenleaf, *Invariant Means on Topological Groups*. Van Nostrand Mathematical Studies, vol. 16 (Van Nostrand-Reinhold Co, New York, 1969)
78. M. Gromov, Hyperbolic groups, in *Essays in Group Theory*. Mathematical Science Research Institute Publications, vol. 8 (Springer, New York, 1987), pp. 75–263
79. M. Gromov, Endomorphism in symbolic algebraic varieties. J. Eur. Math. Soc. **1**(2), 109–197 (1999)
80. K.W. Gruenberg, Residual properties of infinite soluble groups. Proc. Lond. Math. Soc. Third Ser. **7**, 29–62 (1957)
81. A. Guichardet, Tensor products of C^*-algebras. Dokl. Akad. Nauk SSSR **160**, 986–989 (1965)
82. U. Haagerup, C. Winslow, The Effros-Marechal topology in the space of von Neumann algebras. Am. J. Math. **120**, 667–617 (1998)
83. U. Haagerup, C. Winslow, The Effros-Marechal topology in the space of von Neumann algebras, II. J. Funct. Anal. **171**, 401–431 (2000)
84. D. Hadwin, A noncommutative moment problem. Proc. Am. Math. Soc. **129**(6), 1785–1791 (2001)
85. G. Higman, A finitely generated infinite simple group. J. Lond. Math. Soc. **26**, 61–64 (1951)
86. C.J. Hillar, J. Nie, An elementary and constructive solution to hilbert's 17th problem for matrices. Proc. Am. Math. Soc. **136**, 73–76 (2008)
87. V.F.R. Jones, *Von Neumann Algebras*. Course Lecture Notes (2010). Available at http://www.mmas.univ-metz.fr/~gnc/bibliographie/Operator%20Algebras/VonNeumann-Jones.pdf
88. M. Junge, M. Navascues, C. Palazuelos, D. Perez-Garcia, V.B. Scholz, R.F. Werner, Connes' embedding problem and Tsirelson's problem. J. Math. Phys. **52**(012102), 12 pp. (2011)
89. K. Juschenko, S. Popovych, Algebraic reformulation of Connes' embedding problem and the free group algebra. Isr. J. Math. **181**, 305–315 (2011)
90. D.A. Kazhdan, On the connection of the dual space of a group with the structure of its closed subgroups. Akademija Nauk SSSR. Funkcional\cprime nyi Analiz i ego Priloenija **1**, 71–74 (1967)
91. R.V. Kadison, J.R. Ringrose, *Fundamentals of the Theory of Operator Algebras*, vol. I. Graduate Studies in Mathematics, vol. 15 (American Mathematical Society, Providence, 1997) [Elementary theory, Reprint of the 1983 original]
92. R.V. Kadison, J.R. Ringrose, *Fundamentals of the Theory of Operator Algebras*, vol. II. Graduate Studies in Mathematics, vol. 16 (American Mathematical Society, Providence, 1997)

93. I. Kaplansky, *Fields and Rings*, 2nd edn. Chicago Lectures in Mathematics (The University of Chicago Press, Chicago, 1972)
94. A. Kar, N. Nikolov, A non-LEA sofic group (2014). arXiv:1405.1620
95. A.S. Kavruk, Nuclearity related properties in operator systems (2011). Preprint, available at arXiv:1107.2133
96. A.S. Kavruk, The weak expectation property and riesz interpolation (2012). Available at arXiv:1201.5414
97. A.S. Kavruk, V.I. Paulsen, I.G. Todorov, M. Tomforde, Tensor products of operator systems. J. Funct. Anal. **261**(2), 267–299 (2011)
98. D. Kazhdan, Connection of the dual space of a group with the structure of its closed subgroup. Funct. Anal. Appl. **1**, 63–65 (1967)
99. D. Kerr, H. Li, Entropy and the variational principle for actions of sofic groups. Invent. Math. **186**(3), 501–558 (2011)
100. D. Kerr, H. Li, Soficity, amenability, and dynamical entropy. Am. J. Math. **135**(3), 721–761 (2013)
101. E. Kirchberg, On semisplit extensions, tensor products and exactness of group C^*-algebras. Invent. Math. **112**, 449–489 (1993)
102. E. Kirchberg, Discrete groups with Kazhdan's property T and factorization property are residually finite. Math. Ann. **299**, 551–563 (1994)
103. I. Klep, M. Schweighofer, Connes' embedding problem and sums of hermitian squares. Adv. Math. **217**, 1816–1837 (2008)
104. P.H. Kropholler, Baumslag-solitar groups and some other groups of cohomological dimension two. Commentarii Mathematici Helvetici **65**(4), 547–558 (1990)
105. E.C. Lance, On nuclear C^*-algebras. J. Funct. Anal. **12**, 157–176 (1973)
106. E. Lindenstrauss, B. Weiss, Mean topological dimension. Isr. J. Math. **115**(1), 1–24 (2000)
107. T.A. Loring, *Lifting Solutions to Perturbing Problems in C^*-Algebras*. Fields Institute Monographs, vol. 8 (American Mathematical Society, Providence, 1997)
108. J. Łośm, Quelques remarques, théorèmes et problèmes sur les classes définissables d'algèbres, in *Mathematical Interpretation of Formal Systems* (North-Holland, Amsterdam, 1955), pp. 98–113
109. P. Lucke, S. Thomas, Automorphism groups of ultraproducts of finite symmetric groups. Commun. Algebra **39**, 3625–3630 (2011)
110. M. Lupini, Class-preserving automorphisms of universal hyperlinear groups (2013). Available at arXiv:1308.3737
111. M. Lupini, Logic for metric structures and the number of universal sofic and hyperlinear groups. Proc. Am. Math. Soc. **142**(10), 3635–3648 (2013)
112. O. Marechal, Topologie et structure borelienne sur l'ensemble des algebres de von Neumann. Compt. Rend. Acad. Sc. Paris **276**, 847–850 (1973)
113. S. Meskin, Nonresidually finite one-relator groups. Trans. Am. Math. Soc. **164**, 105–114 (1972)
114. R. Moreno, L.M. Rivera, Blocks in cycles and k-commuting permutations (2013). arXiv:1306.5708
115. F.J. Murray, J. von Neumann, On rings of operators, IV. Ann. Math. 2 **44**, 716–808 (1943)
116. H. Neumann, Generalized free products with amalgamated subgroups. Am. J. Math. **70**, 590–625 (1948)
117. D. Ornstein, B. Weiss, Entropy and isomorphism theorems for actions of amenable groups. J. Anal. Math. **48**, 1–141 (1987)
118. N. Ozawa, About the QWEP conjecture. Int. J. Math. **15**(5), 501–530 (2004)
119. N. Ozawa, There is no separable universal II_1-factor. Proc. Am. Math. Soc. **132**(2), 487–490 (electronic) (2004)
120. N. Ozawa, About the Connes embedding conjecture. Jpn. J. Math. **8**(1), 147–183 (2013)
121. N. Ozawa, Tsirelson's problem and asymptotically commuting unitary matrices. J. Math. Phys. **54**(032202), 8 pp. (2013)

122. A.L.T. Paterson, *Amenability*. Mathematical Surveys and Monographs, vol. 29 (American Mathematical Society, Providence, 1988)
123. L. Păunescu, On sofic actions and equivalence relations. J. Funct. Anal. **261**(9), 2461–2485 (2011)
124. L. Păunescu, All automorphisms of the universal sofic group are class preserving (2013). Preprint, available at arXiv:1306.3469
125. V. Pestov, Hyperlinear and sofic groups: a brief guide. Bull. Symb. Log. **14**, 449–480 (2008)
126. V. Pestov, A. Kwiatkowska, *An introduction to Hyperlinear and Sofic Groups*. London Mathematical Society Lecture Notes, vol. 406 (Cambridge University Press, Cambridge, 2012), pp. 145–186
127. S. Popa, Correspondences. INCREST preprint 56 (1986). Available at http://www.math.ucla.edu/~popa/popa-correspondences.pdf
128. R.T. Powers, E. Størmer, Free states of the canonical anticommutation relations. Commun. Math. Phys. **16**, 1–33 (1970)
129. C. Procesi, M. Schacher, A non commutative real Nullstellensatz and Hilbert's 17th problem. Ann. Math. **104**, 395–406 (1976)
130. F. Rădulescu, A non-commutative, analytic version of Hilbert's 17th problem in type II_1 von Neumann algebras, in *Von Neumann Algebras in Sibiu*. Theta Seriers in Advanced Mathematics, vol. 10 (Theta, Bucharest, 2008), pp. 93–101
131. F. Rădulescu, The von Neumann algebra of the non-residually finite Baumslag group $\langle a, b \, | \, ab^3 a^{-1} = b^2 \rangle$ embeds into R^ω, in *Hot Topics in Operator Theory*. Theta Series in Advanced Mathematics (Theta, Bucharest, 2008), pp. 173–185
132. D. Romik, Stirling's approximation for n!: the ultimate short proof? Am. Math. Mon. **107**(6), 556–557 (2000)
133. J.M. Ruiz, On Hilbert's 17th problem and real Nullstellensatz for global analytic functions. Math. Z. **190**, 447–454 (1985)
134. I.E. Segal, Irreducible representations of operator algebras. Bull. Am. Math. Soc. **53**, 73–88 (1947)
135. S. Shelah, *Classification Theory and the Number of Nonisomorphic Models*, 2nd edn. Studies in Logic and the Foundations of Mathematics, vol. 92. (North-Holland, Amsterdam, 1990)
136. S. Shelah, Vive la difference, III. Isr. J. Math. **166**, 61–96 (2008)
137. D. Sherman, Notes on automorphisms of ultrapowers of II_1 factors. Stud. Math. **195**(3), 201–217 (2009)
138. M.H. Stone, Postulates of barycentric calculus. Ann. Mat. Pura Appl. **29**(1), 25–30 (1949)
139. M. Takesaki, On the cross-norm of the direct product of C*-algebras. Tohoku Math. J. **16**, 111–122 (1964)
140. M. Takesaki, *Theory of Operator Algebras I* (Springer, Berlin, 2001)
141. T. Tao, Finite subsets of groups with no finite models. Post in the blog "What's new"(2008). Available at https://terrytao.wordpress.com/2008/10/06/finite-subsets-of-groups-with-no-finite-models/
142. A. Thom, Sofic groups and Diophantine approximation. Commun. Pure Appl. Math. **61**, 1155–1171 (2008)
143. A. Thom, Examples of hyperlinear groups without factorization property. Groups Geom. Dyn. **4**(1), 195–208 (2010)
144. A. Thom, About the metric approximation of Higman's group. J. Group Theory **15**(2), 301–310 (2013)
145. S. Thomas, On the number of universal sofic groups. Proc. Am. Math. Soc. **138**(7), 2585–2590 (2010)
146. T. Turumaru, On the direct-product of operator algebras, II. Tôhoku Math. J. 2 **5**, 1–7 (1953)
147. D. Voiculescu, Symmetries of some reduced free product C*-algebras, in *Operator Algebras and Their Connections with Topology and Ergodic Theory*, ed. by Springer. Lecture Notes in Mathematics, vol. 1132 (Springer, Berlin, 1985), pp. 556–588
148. D. Voiculescu, The analogues of entropy and of Fischer's information measure in free probability theory, II. Invent. Math. **118**, 411–440 (1994)

149. J. von Neumann, Zur allgemeinen theorie des masses. Fundam. Math. **12**, 73–116 (1929)
150. B. Weiss, Sofic groups and dynamical systems. Ergodic theory and harmonic analysis (Mumbai, 1999) Sankhyā Ser. A **62**(3), 350–359 (2000)
151. A. Zalesskiĭ, On a problem of Kaplansky. Dokl. Akad. Nauk SSSR **203**, 477–495 (1972)

Index

algebraic integer, 57
approximate morphism
  of factors, 73, 74
  of invariant length groups, 17, 21, 23, 25,
    40, 43, 75
approximation property, 43
axiomatizable class, 40, 49
  of invariant length groups, 35
  of rank rings, 49
  of tracial von Neumann algebras, 53

Banach space, 105
  dual, 105
  strictly convex, 105, 106
basic formula
  for invariant length groups, 33, 34
  for rank rings, 48
  for tracial von Neumann algebras,
    52
Bernoulli shift, 8
  amenable, 64
  integer, 61
  sofic, 67

C*-algebra, 1, 113
  full, 114
  weakly cp complemented, 86
C*-identity, 113
center, 2, 53
commutant, 2
  double, 2

conjecture
  algebraic eigenvalues, 57
  Connes' embedding, 4–7, 73–75, 78, 80,
    81, 85, 86, 100
  direct finiteness, 10, 50
  Gottschalk's surjunctivity, 8, 10, 61, 65, 67
  idempotent elements, 11
  Kervaire-Laudenbach, 41, 42
  nilpotent elements, 11
  trace of idempotents, 11
  units, 11
  zero divisors, 11
countably saturated
  invariant length group, 37
  rank ring, 49
  tracial von Neumann algebra, 55

definable set, 57

elementary equivalence, 35
elementary property, 35, 53, 54
  of rank rings, 49
  of tracial von Neumann algebras, 53
entropy
  amenable group action, 64
  single transformation, 61
  sofic group action, 66
equivariant function, 8
eventually, 79
extension, 24, 25, 31
  HNN, 25, 27, 29

LECTURE NOTES IN MATHEMATICS  Springer

Edited by J.-M. Morel, B. Teissier; P.K. Maini

Editorial Policy (for Multi-Author Publications: Summer Schools / Intensive Courses)

1. Lecture Notes aim to report new developments in all areas of mathematics and their applications - quickly, informally and at a high level. Mathematical texts analysing new developments in modelling and numerical simulation are welcome. Manuscripts should be reasonably selfcontained and rounded off. Thus they may, and often will, present not only results of the author but also related work by other people. They should provide sufficient motivation, examples and applications. There should also be an introduction making the text comprehensible to a wider audience. This clearly distinguishes Lecture Notes from journal articles or technical reports which normally are very concise. Articles intended for a journal but too long to be accepted by most journals, usually do not have this "lecture notes" character.

2. In general SUMMER SCHOOLS and other similar INTENSIVE COURSES are held to present mathematical topics that are close to the frontiers of recent research to an audience at the beginning or intermediate graduate level, who may want to continue with this area of work, for a thesis or later. This makes demands on the didactic aspects of the presentation. Because the subjects of such schools are advanced, there often exists no textbook, and so ideally, the publication resulting from such a school could be a first approximation to such a textbook. Usually several authors are involved in the writing, so it is not always simple to obtain a unified approach to the presentation.

   For prospective publication in LNM, the resulting manuscript should not be just a collection of course notes, each of which has been developed by an individual author with little or no coordination with the others, and with little or no common concept. The subject matter should dictate the structure of the book, and the authorship of each part or chapter should take secondary importance. Of course the choice of authors is crucial to the quality of the material at the school and in the book, and the intention here is not to belittle their impact, but simply to say that the book should be planned to be written by these authors jointly, and not just assembled as a result of what these authors happen to submit.

   This represents considerable preparatory work (as it is imperative to ensure that the authors know these criteria before they invest work on a manuscript), and also considerable editing work afterwards, to get the book into final shape. Still it is the form that holds the most promise of a successful book that will be used by its intended audience, rather than yet another volume of proceedings for the library shelf.

3. Manuscripts should be submitted either online at www.editorialmanager.com/lnm/ to Springer's mathematics editorial, or to one of the series editors. Volume editors are expected to arrange for the refereeing, to the usual scientific standards, of the individual contributions. If the resulting reports can be forwarded to us (series editors or Springer) this is very helpful. If no reports are forwarded or if other questions remain unclear in respect of homogeneity etc, the series editors may wish to consult external referees for an overall evaluation of the volume. A final decision to publish can be made only on the basis of the complete manuscript; however a preliminary decision can be based on a pre-final or incomplete manuscript. The strict minimum amount of material that will be considered should include a detailed outline describing the planned contents of each chapter.

   Volume editors and authors should be aware that incomplete or insufficiently close to final manuscripts almost always result in longer evaluation times. They should also be aware that parallel submission of their manuscript to another publisher while under consideration for LNM will in general lead to immediate rejection.

4. Manuscripts should in general be submitted in English. Final manuscripts should contain at least 100 pages of mathematical text and should always include

   – a general table of contents;
   – an informative introduction, with adequate motivation and perhaps some historical remarks: it should be accessible to a reader not intimately familiar with the topic treated;
   – a global subject index: as a rule this is genuinely helpful for the reader.

   Lecture Notes volumes are, as a rule, printed digitally from the authors' files. We strongly recommend that all contributions in a volume be written in the same LaTeX version, preferably LaTeX2e. To ensure best results, authors are asked to use the LaTeX2e style files available from Springer's web-server at
   ftp://ftp.springer.de/pub/tex/latex/svmonot1/ (for monographs) and
   ftp://ftp.springer.de/pub/tex/latex/svmultt1/ (for summer schools/tutorials).
   Additional technical instructions, if necessary, are available on request from:
   lnm@springer.com.

5. Careful preparation of the manuscripts will help keep production time short besides ensuring satisfactory appearance of the finished book in print and online. After acceptance of the manuscript authors will be asked to prepare the final LaTeX source files and also the corresponding dvi-, pdf- or zipped ps-file. The LaTeX source files are essential for producing the full-text online version of the book. For the existing online volumes of LNM see:
   http://www.springerlink.com/openurl.asp?genre=journal&issn=0075-8434.
   The actual production of a Lecture Notes volume takes approximately 12 weeks.

6. Volume editors receive a total of 50 free copies of their volume to be shared with the authors, but no royalties. They and the authors are entitled to a discount of 33.3 % on the price of Springer books purchased for their personal use, if ordering directly from Springer.

7. Commitment to publish is made by letter of intent rather than by signing a formal contract. Springer-Verlag secures the copyright for each volume. Authors are free to reuse material contained in their LNM volumes in later publications: a brief written (or e-mail) request for formal permission is sufficient.

Addresses:
Professor J.-M. Morel, CMLA,
École Normale Supérieure de Cachan,
61 Avenue du Président Wilson, 94235 Cachan Cedex, France
E-mail: morel@cmla.ens-cachan.fr

Professor B. Teissier, Institut Mathématique de Jussieu,
UMR 7586 du CNRS, Équipe "Géométrie et Dynamique",
175 rue du Chevaleret,
75013 Paris, France
E-mail: teissier@math.jussieu.fr

For the "Mathematical Biosciences Subseries" of LNM:

Professor P. K. Maini, Center for Mathematical Biology,
Mathematical Institute, 24-29 St Giles,
Oxford OX1 3LP, UK
E-mail: maini@maths.ox.ac.uk

Springer, Mathematics Editorial I,
Tiergartenstr. 17,
69121 Heidelberg, Germany,
Tel.: +49 (6221) 4876-8259
Fax: +49 (6221) 4876-8259
E-mail: lnm@springer.com

Printed in the United States
By Bookmasters